Wounds of War

BY HERBERT HENDIN

The Age of Sensation

Black Suicide

Psychoanalysis and Social Research

Suicide and Scandinavia

Suicide in America

WITH ANN POLLINGER HAAS

Adolescent Marijuana Abusers and Their Families

Chronic Marijuana Use in Adults

Wounds of War

WOUNDS
OF WAR

The Psychological Aftermath
of Combat in Vietnam

HERBERT HENDIN

ANN POLLINGER HAAS

Basic Books, Inc., Publishers   *New York*

Library of Congress Cataloging in Publication Data

Hendin, Herbert, 1926–
  Wounds of war.

  References: p. 245
  Includes index.
    1. Veterans—United States—Psychology
2. Vietnamese Conflict, 1961–1975—United States.
3. Vietnamese Conflict, 1961–1965—Psychological
aspects.   4. Posttraumatic stress disorder—United
States.   I.   Haas, Ann Pollinger, 1944–
II. Title.
UB369.H46   1984   616.85′212   83–46074
ISBN 0–465–09259–4

Contents

Preface

THIS BOOK was written about the Vietnam experience and its effects on the men who fought there because we believe it has something to say that has not been told, and cannot be told, by anyone who has not come to know well men who saw extensive combat in that war. Large-scale studies of Vietnam veterans, relying on questionnaires or single interview procedures, have reported numerous postwar adjustment problems of men whose lives were interrupted by service in Vietnam, but have provided only a glimpse of the profound personal transformation that occurs in most who were exposed to heavy combat. Much of the clinical literature as well, fails to capture adequately the intensely personal, subjective experience of combat for individual veterans, nor has it illuminated how that experience continues to shape the rest of their lives.

At the heart of the Vietnam experience was the sustained exposure to life-threatening combat, of which veterans' written accounts—fictional and non-fictional—have begun to emerge only in recent years. Combat in Vietnam was so overpowering and shattering an experience that it has required a considerable period of time for it to be mastered sufficiently for participants to write about it. No doubt for comparable reasons, Erich Maria Remarque's *All Quiet on the Western Front,* perhaps the finest novel ever written about combat, appeared eleven years after the end of World War I. Although many of the recent writings by combat veterans have added a great deal to our understanding of the Vietnam War, what is often missing in their books, as in those which followed earlier wars, is an understanding of the meaning of the experience for the veteran, and the ways in which it intrudes upon, alters, and often dominates his postwar life.

The uninitiated reader may wonder how or why the experience of the combat soldier in Vietnam and afterward differed from that of American soldiers in other wars. Although no systematic comparison of the Vietnam experience with that of previous wars is possible (nor is it our goal), we hope that by the end of this book the reader will have a conception of the important similarities as well as the striking differences between the experiences of American soldiers in Vietnam and their earlier American counterparts.

In 1978, our research and clinical work at the Franklin Delano Roosevelt Veterans Administration Hospital in Montrose, New York, first brought us in contact with Vietnam combat veterans who were suffering from the aftermath of their combat experiences. These men had developed a coherent set of symptoms known as a "posttraumatic stress disorder." The lives of some of them had been indelibly altered by the disorder, while in others the symptoms were present in less incapacitating forms. Almost none of the men, we found, had ever spoken intimately to anyone of combat, nor were many able to relate difficulties in work, relationships, or with drug and alcohol abuse—all of which clearly originated in combat—to their experiences in Vietnam.

During the last five years, we have worked with over a hundred Vietnam combat veterans with varying degrees and manifestations of posttraumatic stress. We have also worked with a smaller, specially selected group of men who, although participants in intense, sustained combat, had never developed a stress disorder.

Each of these veterans went through a systematic evaluation procedure, consisting of a lengthy written questionnaire and a series of five semistructured clinical interviews. The evaluation sought to uncover not merely what the veteran experienced in combat, but the meaning that the experience had and was continuing to have for him. In addition, the procedure aimed to establish the presence or absence of posttraumatic stress, to delineate the particular form the disorder took, and, where appropriate, to propose a treatment plan for the veteran. Many of the men who completed the evaluation were treated in individual stress-oriented psychotherapy for periods ranging from several months to more than two years.

Although from the outset our intention has been to work toward more effective ways of conceptualizing and treating posttraumatic stress disorder, rather than to generalize about Vietnam or the men who fought there, the veterans we have worked with do, in fact, mirror the wider population of Vietnam veterans in terms of demographic and social background characteristics. Vietnam, in contrast to past wars in which American forces have participated, has been characterized as a "working-class war." The socioeconomic backgrounds of Vietnam veterans underscore the accuracy of that characterization: seven out of every ten men who served in Vietnam came from families in which the main wage earner held a blue-collar position, in contrast to only half of their non-veteran peers.* An almost

*The statistical data on the overall population of Vietnam veterans and their non-veteran peers have been taken from Egendorf, A., Kadu-

identical proportion of the veterans we have worked with are from similar working-class backgrounds.

Among our group, as among Vietnam veterans nationwide, approximately the same number came from large urban centers, small towns, and rural areas. About three-quarters of both our veterans and Vietnam veterans in general were from intact families, and almost half had fathers with combat experience in earlier wars. As a group, those who fought in Vietnam were younger than veterans of earlier wars, with an average age at the time of their Vietnam experience of only 19.7 years.

Against this background, our group of over a hundred veterans may be regarded as a microcosm within which the impact of the Vietnam experience can fruitfully be explored and understood. Coming to know these men so well, understanding what Vietnam meant to them, and following their struggles to resolve their painful ties to the war, have given us a unique perspective on the sources, manifestations, and complications of post-traumatic stress. In particular, we wish to share in this book what we have learned about how the lingering stress associated with wartime experiences can be treated.

We brought to this work the very different disciplines of sociology and psychodynamic psychiatry. We are colleagues in a research center where the integration of these disciplines has been found to be productive in the exploration of a variety of problems. This was particularly true in the present study: while one of us evaluated

---

shin, C., Laufer, R.S., Rothbart, G., and Sloan, L., *Legacies of Vietnam: Comparative Adjustment of Veterans and Their Peers.* Washington, D.C.: U.S. Govt. Printing Office, 1981. This study, prepared for the Veterans Administration, used national probability sampling techniques that resulted in a comprehensive description of those who went to war in Vietnam contrasted to those who did not.

and treated the veterans whose psychotherapy is described here, and the other developed and administrated our questionnaire, we mutually discussed and analyzed the material from both sources. Moreover, much was learned from the discussion of cases evaluated and treated by our other colleagues at the Center for Psychosocial Studies.

In most of the chapters that follow we have relied on the case history format in order to emphasize the uniqueness of each of the many veterans who, by sharing their experiences with us, have made this book possible. Although names and some specifically identifying features have been changed, we have tried to convey the essence of the precombat, combat, and postcombat experiences of these men.

Work with veterans of the Vietnam War has increased our knowledge of traumatic stress in general, and that resulting from this war in particular. Insofar as their suffering increases awareness of the human price of war, their sacrifices will be not so much in vain as many of them fear. It is our hope that this book will serve as a source of greater understanding of the wounds of war, and as a guide for all—family, friends, and professional therapists—who will play a part in healing those wounds.

Acknowledgments

WE ARE INDEBTED to our colleagues at the Center for Psychosocial Studies who discussed much of the material in this book with us and provided their valuable insights. Paul Singer and Frank Gold, who is also Chief of Psychiatry at the Veterans Administration Hospital in Montrose, New York, deserve particular mention in this regard. William Houghton, also a colleague at the Center, read the manuscript and shared a number of ideas which we found helpful. In addition, we are grateful for the contribution of a former colleague, Richard Ulman, who collected some of the background information on three veterans whose cases are presented in chapters 3 and 9.

Alfred Freedman, Chairman of the Department of Psychiatry of New York Medical College, was a source of continual support and encouragement. Corydon Heard, Director of the Veterans Administration Hospital in Montrose (now retired) facilitated our work with Vietnam veterans at that hospital.

A number of other individuals read the manuscript and made helpful suggestions for its improvement. Josephine Hendin deserves our special thanks for her review of many different drafts of the material. We are also grateful for the suggestions made by Lawrence Kolb, Distinguished Physician in Psychiatry for the Veterans Administration, and Marshall Carter, who served in Vietnam as an officer in the U.S. Marine Corps.

Judith Greissman, our editor, was both incisively critical and generously supportive in the writing and rewriting of this manuscript, and we are deeply appreciative of her efforts.

We are grateful to Monica DeRonda, Susan Brownell, and Valerie Hewitt, who typed most of the original case material, and to Margaret Williams, who typed the manuscript.

Our work on posttraumatic stress was partially supported by a grant from the National Institute of Drug Abuse (DA03361). Many people at the Institute were of help to us, but we would like to acknowledge in particular Dan Lettieri, currently with the National Institute on Alcohol Abuse and Alcoholism, and Meyer Glantz, our project officer.

Most of all we wish to acknowledge the Vietnam veterans who shared their experiences with us.

Trauma

1

Wounds of War

IN no prior war fought by the United States has the actual combat experience of our fighting men been less understood by the public than in the Vietnam War. Even when the country became aware that the men who fought in Vietnam had not been treated fairly while they were there or afterward, the impulse to make restitution was not accompanied by a need to know about or understand their experience.

Although the country in general and the media in particular are aware that psychological difficulties were a frequent consequence of exposure to the trauma of life-threatening combat events in Vietnam, the awareness has been of the sort that trivializes rather than comprehends the experience. The after-reaction to the stress of combat, popularly called "delayed stress," has become a pat explanation for any strange or violent behavior by a Vietnam veteran, with no genuine attempt to understand the individual or his experience.

The reaction to combat captures the essence of the experience and what it meant to the Vietnam veteran. He may have been involved in extensive combat; much of what he recounts may seem a possible and sufficient basis for a traumatic reaction. But the veteran's recurrent nightmares and experiences in which he relives what happened to him in combat tell us in a way he often cannot otherwise express what was most traumatic for him, what combat signified for him, and what in the experience he cannot resolve. To understand one veteran and the interrelation of his combat experiences, their meaning for him, and his postwar difficulties is to begin to understand the Vietnam War experience in a way that few Americans have.

To a degree unparalleled in our earlier wars, combat in Vietnam involved the killing of women, children, and the elderly: some of whom were armed fighters, some of whom were killed inadvertently, and some of whom were killed in retaliation for deaths caused by their countrymen. Regardless of the circumstances, and even in situations where the veteran had to kill to save his life, guilt over such killing is profoundly disturbing to most veterans and plays, in comparison with other wars, a far more significant role in the stress disorders of Vietnam veterans.

A veteran who has come to accept most of his harrowing combat experiences is still tormented in his nightmares by a memory of a night attack in which he shot a Vietcong woman who was firing at him, only to discover in the morning that he had killed a baby strapped to her back. A veteran who saw extensive combat and was seriously wounded is haunted by one death—that of an old man who threw a grenade at his squad. He says sadly, "The man should have died in bed with his family, not be killed by a kid of eighteen like me." These veterans find it hard to tell others their experiences. When they do, many do not want to listen.

Some Americans who opposed the war find it difficult to empathize with the young men who went to Vietnam. Some who supported the war find it hard to look at its corrosive impact on those who fought it. Even veterans of World War II often do not want to hear what their sons' combat experiences in Vietnam actually were.

Much has been said of the excesses of American servicemen in Vietnam, with little awareness of the context in which they occurred. Little, however, is said of the common occurrence that is seared in the memory of many combat soldiers: a captured American tied to a tree, his genitals cut off and stuffed in his mouth. The sight of the mutilated body, intended by the Vietcong to create fear in our soldiers, was at times used by American officers to create rage and ferocity in them instead.

To Americans, the horror of the war for the Vietnamese appears to be somewhat better understood than the destructive impact that it had on the men we sent to fight. If the Vietnam War is unique in the degree to which Americans have not wished to hear about or understand the experiences of the men who fought, the reluctance to be concerned with the postwar consequences of combat for the veteran has a much longer history. Society has been slow to recognize the ongoing price the soldier pays for participating in combat.

When some of the symptoms of what we now call a posttraumatic stress disorder—nightmares, insomnia, an excessive startle reaction to loud noise, and outbursts of anger—were first recognized among hospitalized veterans of World War I, the condition was assumed to be relatively rare. If we pitied "shell-shocked" combat survivors we also judged them to be an unfortunate minority who paid a high price for their participation in the war. Our need to see our soldiers as heroes in war and to forget them in peace, and our tendency to regard their postwar difficulties as weaknesses interfering with our idealized picture of them, caused us not to notice that even heroes pay a high price for their wartime actions. Only since World War II have we begun to realize that killing, sustained exposure to the possibility of sudden death, and witnessing the violent deaths of friends have lasting traumatic consequences for a high percentage of combat soldiers.

Audie Murphy, the most decorated American hero of World War II, spent two years with the infantry in Southern Europe, during which he was wounded three times, and credited with killing 240 German soldiers. When he died in 1971, it was reported that since the war he had been plagued by nightmares in which he relived his combat experiences, and was unable to sleep without a loaded German Walther automatic pistol under his pillow. How his waking life was also marked by preoccupation with his combat experiences was suggested by his response to an interviewer's question about how combat soldiers manage to survive a war: "I don't think they ever do," he replied.[1] Audie Murphy had developed many of the posttraumatic stress symptoms that we are now commonly seeing in Vietnam veterans.

The American involvement in Vietnam may be the crucible we needed to at long last explore the corrosive effects of war on the combat soldier. Over the past fifteen years we have seen increasing numbers of Vietnam combat veterans suffering from stress disorders engendered by the trauma of combat. It has been estimated by the Veterans Administration that about half of the soldiers who saw combat in Vietnam are afflicted with posttraumatic stress.[2] Over half a million combat veterans are probably involved. But the impact of their behavior on their wives, children, parents, employers, and co-workers brings to many millions the numbers affected.

The one-year tour of duty for our armed forces in Vietnam was partly intended to insure against the development of incapacitating stress reactions in our soldiers during combat. In that purpose, the policy succeeded to some extent,[3] while insuring that more veterans would experience stress symptoms after they returned home. Combat in Vietnam heightened our awareness of the frequent delay between traumatic events and the development of stress, which results from the soldier's need to suspend and control the overwhelming fear and rage stimulated by life-threatening experiences. The delayed stress reaction is, in a sense, the price he later pays for suspension of his emotions in the service of effective combat.

The political divisiveness surrounding the Vietnam War, and the lack of support for returning soldiers, have been suggested as primary factors contributing to the widespread problems of Vietnam veterans.[4] In our experience with veterans, however, the nature of the trauma of combat, and of combat in Vietnam in particular, appears to have been more critical in the development of posttraumatic stress disorders than the issue of whether or not we should have been involved in Vietnam. Most of the men who fought accepted participation as their duty, and a sizable number supported the United States involvement in the war.[5] Many still feel that way and are critical of their government primarily for not more vigorously pursuing the war to a successful conclusion. Many others came to see the war as destructive for all concerned and without clear purpose. Post-traumatic stress disorders, however, are distributed among combat veterans of both persuasions.

Some of those who developed severe stress reactions, whom we will describe in the chapters that follow, were honored as returning heroes in their communities. But more are bitter at the hostility or indifference with which they were greeted on their return. Certainly the lack of appreciation experienced by these men as they returned from an unpopular war contributed to their difficulties, but not as much as what they had experienced in combat. Audie Murphy developed a post-traumatic stress disorder even though he was celebrated as a hero of a war the country supported.

The particular quality of combat in Vietnam was crucial to the genesis of posttraumatic stress. The close, personal encounter with death and with killing is at the heart of the disorder. In Vietnam the ambushes, the search-and-destroy missions, and the nature of guerrilla warfare, with its sudden unexpected combat engagements and its lack of clarity as to who was the enemy, gave these encounters a high frequency. The relative youth of the average American combatant in Vietnam (under twenty-years-old, compared to an average age of twenty-six for World War II soldiers, and twenty-four for the Korean War) and the lack of *esprit de corps* among soldiers resulting from the one-year rotation in and out of the country (as individuals, not as units) heightened the stress of warfare in Vietnam.[6]

To the extent that social unrest over the war influenced the decision to fight in a limited way, thus affecting the kind of war that was fought, it contributed to the environment of stress. Military strategy in Vietnam, epitomized by the demand for body counts as an index of success, and the constant taking, leaving, and retaking of the same terrain, fostered the sense among soldiers in Vietnam that the war was waged foolishly, that the American government's commitment was less than wholehearted, and that their lives and those of their friends were being wasted. Feelings of alienation and mistrust toward the government and the country they fought for were greater than in veterans of past wars. The personal pain many Vietnam veterans feel from their combat experiences is aggravated by a continuing sense that their sacrifices were unappreciated, and a continuing suspicion that they were unnecessary.

This book is an outgrowth of our attempt to understand how posttraumatic stress disorders have developed out of the complexities of individual, social, and cultural factors surrounding the Vietnam War. To do this it is first necessary to examine the evidence that the disorder existed, unrecognized, in combat veterans long before World War I; to trace how our conception of this disorder evolved; and to indicate how reactions to the trauma of combat are different from those in civilian disasters. We are then in a position to understand how, among all wars, the Vietnam experience and the reactions to it were unique. These themes are the subject of chapter 2, although the special quality of the Vietnam experience is, in a sense, the subject of this entire book.

The conception of posttraumatic stress that emerged from World War I changed and expanded during and after World War II. The realization of certain similarities between combat veterans, concentration camp survivors, Japanese survivors of the atomic explosions in Hiroshima and Nagasaki, and the victims of floods and other civilian disasters contributed to the expansion of our knowledge.

Yet combat is different from other forms of trauma. In

contrast to the civilian victim, the combat soldier is usually an active participant in the life-threatening and life-destroying action. Not only is killing a key goal of the combat mission, but firing a weapon in an attempt to kill the enemy often serves as a way of relieving the soldier's fear of being killed. Killing and its use in controlling one's fear have consequences that give a different quality to the posttraumatic stress disorders of combat veterans.

Chapter 3 presents the core of our approach to understanding posttraumatic stress disorders among Vietnam veterans. This approach rests on the recognition that while the traumatic experience of combat is at the heart of the disorder, neither the subjective perceptions of combat, nor the subsequent reactions to it, are the same for all veterans. The unique personal and social characteristics that each individual brought to combat played a role in shaping his combat experiences, in influencing his perceptions of traumatic combat events, and in determining the specific meanings that such events had, and continue to have, for him. Postcombat adaptation, as well as precombat factors, contribute to helping us understand the meanings of combat for the veteran.

Understanding what combat has meant to the individual provides the basis for treating posttraumatic stress disorders among veterans. Chapter 4 presents an introduction to stress-oriented psychotherapy. The veterans we have treated, some of whose cases are presented in chapters 5, 6, and 7, provide the most powerful and revealing evidence of what combat meant to these men. These chapters also illustrate how therapy focused on understanding the meanings of combat can help the individual integrate his wartime experiences with the rest of his life.

Guilt, and the effort to protect oneself against it, play a special role in the posttraumatic stress disorders of Vietnam veterans. Traditionally, soldiers have developed guilt because they were afraid or ineffective, because they killed when they did not have to, or simply because they killed. The moral ambiguity surrounding who should or should not be killed in Vietnam and the breakdown of the codes of conduct, which at least to some degree governed behavior in other wars, created an inordinate number of guilt-generating situations.

Chapter 5 explores various sources which an emotion like guilt can have. For Dennis Allen, one veteran described in this chapter, guilt centered around his detached curiosity toward the horror around him. Another veteran, Tom Bradley, who was a witness to a rape-murder and to the killing of many civilians, believed that what he saw was understandable and claimed to feel neither responsibility nor guilt. His nightmares contradicted this attitude and were therapeutically invaluable in helping him to understand and come to terms with his feelings. A third man, Bill Clark, was all too conscious of actions he committed while feeling out of control during his combat tour. He was not aware, however, of his need for punishment, which was reflected in over a dozen serious car accidents, self-destructive behavior at work and at home, and a recurrent nightmare in which he would see incoming shells with his name written on them.

Two major ways in which veterans deal with posttraumatic stress are the focus of chapters 6 and 7. In one adaptation veterans perceive civilian life as an extension of the war and attempt to protect themselves in ways more appropriate to the battlefield. This "paranoid" manifestation of the stress disorder is characterized by suspicion and mistrust, hypersensitivity and a readiness to fight, and a cold, unemotional way of relating to people. Frank O'Donnell, the veteran described in chapter 6, developed the ability to react aggressively and effectively when confronted with life-threatening combat situations. His behavior, which in Vietnam was often heroic, had a vastly different character when the stimulus was a motorist who cut in front of his car, or a rude salesperson who aroused fantasies of killing in him.

The other adaptation is "depressive" in nature. The characteristic lack of pleasure and loss of vitality are centered around a state of mourning over combat experiences. Veterans with this adaptation mourn for friends who have died, but they also mourn for what they have lost of themselves, and often perceive themselves as also having died in combat. Such perceptions are reflected in a recurrent nightmare of Jack Edwards (a veteran presented in chapter 7), in which he is turning over dead bodies in Vietnam and discovers his own rotting corpse.

Public concern with Vietnam veterans is often focused on such problems as violence and crime, suicide, and drug and alcohol abuse. There is evidence that each of these problems is more frequent among combat veterans than among either noncombat veterans or nonveterans of the same age. In a high percentage of cases, there is a relationship between asocial or antisocial behavior and combat experiences and these relationships are the subject of chapters 8, 9, and 10.

Although some veterans who have engaged in criminal or violent behavior since the war showed evidence of antisocial behavior prior to service, and others engaged in such behavior while overseas, still others had no such history prior to returning from Vietnam. Warren Saunders, a veteran discussed in chapter 8, came from a warm supportive family, had made a good marriage, and was a devoted father before his nightmarish combat experiences in Vietnam. Severe, untreated posttraumatic stress precipitated the breakup of his family, an addiction to heroin, and a series of bank robberies with other Vietnam veterans for which he was eventually imprisoned.

None of the suicidal combat veterans we have seen to date showed evidence of suicidal behavior prior to combat, and all experienced severe posttraumatic stress symptoms prior to their becoming suicidal. Yet most of these veterans were not in touch with the relationship between their suicidal behavior and their combat experiences. Tony Marco, who is discussed in chapter 9, was satisfied in his marriage and family life and had held a good job for eight years. He recurrently dreamed that he was back in Vietnam and someone would sneak up behind him and cut his throat. Periodically he would be overcome with a desire to end his life and had made several serious suicide attempts. He had no idea why he sometimes became suicidal, nor did he connect his suicidal behavior to his dreams. Uncovering the linkage between his Vietnam experiences, the meaning of these experiences, and his postwar behavior had a dramatic impact on his life.

The vast majority of combat veterans were not substance abusers prior to combat. Some began heavy use of drugs or alcohol to relieve their fears during combat in Vietnam. For others drug abuse began with medical treatment of wounds sustained in combat. And for still others, drug and alcohol abuse began only after returning from the war, often as a direct response to the symptoms of posttraumatic stress. It is striking how men like Steve Wallace, one of the veterans discussed in chapter 10, find sleep free of combat nightmares possible only with drugs. Equally striking is the way in which substance abuse often stops with the alleviation of stress symptoms.

Are there individuals who are relatively invulnerable to posttraumatic stress? If so, what protects them? Debate over such opposite positions as "everyone has his breaking point," and "only the neurotic soldier will break down," has taken place without an attempt to study soldiers who have been through intensive combat and have not developed lasting symptoms. In chapter II we discuss what we have learned from our work with "invulnerable" veterans. Here, emphasis is placed on identifying factors related to a type of combat adaptation that seems to have protected some from the lasting scars of war.

In the concluding chapter the lessons from the past and present are brought together in an exploration of what our policy toward soldiers in and after combat ought to be, as well as the implications that policy holds for civilians exposed to overwhelming disasters. Finally, we discuss some of the wounds that have been inflicted on the country as a whole by the Vietnam experience, pointing to the need for societal, as well as individual, resolutions of its traumatic effects.

2

Changing Perspectives

PERHAPS the most frequent question raised by the psychological aftermath of Vietnam is why stress responses to combat should be more pervasive after this particular war. Although most clinicians who have worked with veterans since World War II share the opinion that there has been a singularly high incidence of stress disorders after Vietnam, that judgment is confounded by our greater understanding of and greater ability to diagnose the disorder.

If intimate experience with violent death is the source of psychological disorders following combat, these disorders must have developed in soldiers engaged in intense combat in wars prior to this century. Although it was not until World War I that the symptoms were clinically recognized and attention paid to the long-term psychological impact of combat, there was, we believe, evidence of the disorder long before that.

Awareness of acute short-term reactions to combat

was reflected in the *nostalgie,* or homesickness, described among French troops in the Napoleonic wars. Similarly, "irritable heart" was a manifestation of anxiety described in Civil War soldiers. The clinical interest in such reactions, however, centered on weeding out those not fit for combat. It is only a partial oversimplification to say that, before World War I, both military men and clinicians perceived combat as simply a question of courage versus cowardice, with no sense that there might be a lasting price to be paid even for bravery.[1]

Despite the prevailing lack of clinical interest in the subject, historical accounts of the Civil War provide a wealth of largely unexplored evidence that posttraumatic stress afflicted veterans. One dramatic example is found in the life of Lewis Paine, one of the conspirators involved in the assassination of President Abraham Lincoln.[2] Paine, a hardworking, religious youngster adored by his family, joined the Confederate Forces at sixteen, along with two of his older brothers. For the next several years he saw extensive combat throughout the South. One of his brothers was killed, the other permanently disabled, and Paine himself was wounded and captured at Gettysburg. He was placed in a Union hospital, subsequently escaped, and in late 1863 rejoined the Confederate Army in Virginia.

Once considered easy-going, kind, and tenderhearted, Paine became obsessed with killing and death, frequently boasting that he never left wounded enemy soldiers alive and using the skull of a Yankee soldier as an ashtray. Later in the war, he began displaying an explosive temper. Several accounts exist, for example, of his savage beating of a hotel's female servant who did not respond quickly enough to his requests. Although the record is unclear on this point, some evidence exists that he deserted his unit in early 1865. Around this time he began using a variety of aliases and broke off all contact with his family.

When he was twenty, Paine was recruited by John Wilkes Booth to join a conspiracy to capture Lincoln and hold him for ransom in exchange for Confederate prisoners needed to reinforce the South's dwindling forces. There is little doubt that the quality which most attracted Booth to young Paine was his capacity to inflict brutal punishment on anyone who opposed him. As the conspiracy plot changed from one of capturing Lincoln to killing him and his closest advisors, Paine was assigned the task of assassinating Secretary of State William Seward in his home. At the same time, Booth would be killing Lincoln at the Ford Theatre.

Paine's mission was not successfully carried out because Seward was wearing a neck brace (the result of a carriage accident) that partially protected him from the slashes of Paine's knife. But in addition to seriously wounding Seward, Paine beat and slashed two sons and several other household members and killed a servant who tried to intervene before he ran from the house shouting, "I am mad! I am mad!"

After his capture, Paine attempted to commit suicide. At his trial his lawyer pleaded that Paine had become mentally ill as a result of the traumas he had experienced during his almost four years as a combat soldier. Without clinical support for the argument of war-induced "homicidal mania," and with the entire country irate over Lincoln's assassination and the violence of the attack on Seward, the jury was not swayed. Paine was convicted and put to death by hanging.

While historical documentation of cases such as this strongly suggest the unrecognized presence of post-traumatic stress among participants in early wars, soldiers' diaries and autobiographical accounts, as well as the work of authors who used war as the context for their writing, are an additional source of largely unexamined evidence. In the life and work of the notable nineteenth-century American author Ambrose Bierce,[3] who fought in and wrote about the Civil War, we find a persuasively documented expression of the after-effects of combat in an individual with a posttraumatic stress disorder.

Born in 1842, the tenth child of a religious Ohio family, Bierce grew up on a series of poor Indiana farms. His father, a frustrated scholar, preferred learning to farming and owned a large collection of books, and from him Bierce seems to have gotten his love of reading. A sociable adolescent, Bierce was courting a young woman when the Civil War broke out in 1861. He enlisted as a private in the 9th Regiment of Indiana Volunteers.

Serving for the duration of the war, Bierce saw action in the battles of Shiloh, Chickamauga, and Chattanooga. He twice rescued wounded companions at grave risk to his own safety, and was twice wounded himself, once in the head during the Battle of Kenesaw Mountain in Georgia. His merit as a soldier was recognized with a promotion to sergeant and later to lieutenant.

Leaving the military at twenty-four, he took up journalism, for a career that spanned almost fifty years as a columnist, satirist, and political commentator. Much of this time he was associated with the Hearst newspapers, in San Francisco and later in Washington. His best known collection of short stories, *Tales of Soldiers and Civilians,* based on his Civil War experiences, was published in 1891. The brutal recall of detail, preoccupation with death, and descriptions of blood, maiming, and destruction in these stories suggest that Bierce's post-war life was marked by ongoing, intense images of horror.

Bierce's tales reflect a number of themes that have been found to be pervasive in the fantasies of modern-day combat veterans with posttraumatic stress disorders. "An Occurrence at Owl Creek," for example, chronicles the details of a man's wartime escape from certain death. Only at the end of the story do we learn that the escape was the man's final fantasy; that he has, in fact, been killed. This theme has a striking parallel among many Vietnam veterans with posttraumatic stress whose words or lives express a sense that their survival of combat is an illusion and that they have in reality died in the war.

In another Bierce story, "Chickamauga," a young child is in the woods, playing at war, when he falls asleep. A battle has been fought while he was asleep, and he awakens to a surrealistic procession of dead and wounded soldiers. The face of one man that "lacked a lower jaw" is graphically described: "from the upper teeth to the throat was a red gap fringed with hanging shreds of flesh and splinters of bone." The boy leads the mutilated group home to discover it burning and his mother dead, her brains oozing from a small hole in her temple. The child makes a series of inarticulate cries, and we learn he is a deaf mute. The dreamlike return of dead soldiers, the blurring between combat reality and nightmare, and the muteness in the face of ultimate horror are themes which frequently pervade the lives of Vietnam combat veterans.

The detailed realism of Bierce's many accounts of ir-rational, inescapable horrors suggests that his own nightmares were an important source of his writings. There is also considerable evidence that he suffered from "emotional numbing," a symptom that was later recognized as an essential constituent of the post-traumatic stress disorder. Throughout his postwar life, Bierce remained aloof, even in his closest relationships, and reticent about his feelings. In San Francisco, he married a woman named Molly Day, but the marriage was not successful. After the birth of two sons and a daughter, Molly left Bierce. He was never again close to a woman. Relationships with his children were likewise characterized by failure and tragedy: his older son was killed in a shooting brawl over a woman; his younger son died of alcoholism in 1901.

Bierce also suffered from persistent insomnia, a common posttraumatic stress symptom. As a result, he often began writing at midnight, not to go to bed until dawn. He drank heavily, and had frequent outbursts of temper. As a writer for Hearst, he would often resign if a piece had been excessively edited, then have to be calmed down by Hearst himself. Like many of today's stressed combat veterans, Bierce carried a revolver and frequently threatened to use it, although no actual incidents are reported. Once, however, he broke a cane over the head of a collaborator, during an argument over rights of authorship.

Bierce's brother, Albert, saw the dark side of Bierce's personality as the result of his war experiences, but in keeping with the view of the times, attributed the change in his behavior to the head wound Bierce received at Kenesaw Mountain. George Sterling, a pupil of Bierce's, cites Albert's claim that after this incident his brother's "entire nature underwent a change."[4]

Bierce's death reflected the extreme isolation and preoccupation with the themes of violence and war which characterized his life. In 1913, at the age of seventy-one, he moved to Mexico and joined the revolution under Pancho Villa. Although the details are unclear, it appears that after a short time he deserted to the Constitutionalists and was later captured and shot. Sterling describes Bierce's bizarre death as evidence of his long-standing "wish to be slain in war." His ironic bitterness toward the country he served, as reflected in the following lines,[5] suggests an alienation not at all unlike that frequently seen in today's veterans:

> "My country 'tis of thee
> Sweet land of felony
> Of thee I sing . . ."

The evidence of posttraumatic stress in combat veterans remained unrecognized. When the disorder was first described, it was among the victims of civilian traumas ranging from railway accidents to being struck by lightning. The disorder was considered a consequence of physical injury to the nervous system, carrying the names of the various agents thought to have provoked it, for example, "lightning neurosis" or "railway spine."[6] The common use of the term "shell shock" to describe combat reactions among soldiers in World War I was a natural extension of this prevailing view. Although personality changes involving alternations between apathy and over-excitability were also observed in trauma victims, these responses were seen as secondary to physical injury.

Early in his career Sigmund Freud recognized the similarity of the way in which symptoms developed in what was then called traumatic neurosis and hysteria. In both disorders, he found that strong emotion associated with a traumatic situation had been repressed, with subsequent development of symptoms that were psychologically linked to the original trauma.[7] During

21

the early part of World War I, Freud's colleagues believed they had confirmed his views in their observations of the traumatic reactions of soldiers to combat. The predominance of neuropsychiatric opinion, however, continued to favor the view that the disorder was primarily physical.

By the end of World War I, when it was evident that the symptoms of traumatic war neurosis were seen in combat soldiers who had experienced no physical trauma, there began a growing acceptance of the psychological basis of the disorder. In a symposium on the problem held at a 1918 psychoanalytic congress in Budapest, papers presented by Sandor Ferenczi, Karl Abraham, and Ernst Simmel conveyed, somewhat triumphantly, that the experience of the war had confirmed the psychogenic basis of the traumatic neurosis. Ferenczi's introductory paper reflected that tone: "We cannot spare neurologists the reproach of having so long disregarded the pioneer researches of Breuer and Freud concerning the psychical determinations of many nervous disturbances, and of having required the terrible experience of the war to set them right in this respect. When we read in the more recent literature on the subject of the ideas and views which have become so familiar—abreactions, unconscious psychic mechanisms, separation of the affect from the idea, etc.—we might easily imagine ourselves to be in a circle of psychoanalysts . . . ."[8]

If the papers were gloating in tone toward neurologists, there was in all of them an equally pronounced defensive tone in response to psychiatric criticism that the development of war neurosis provided a refutation of Freud's theories concerning the ubiquitous sexual origin of neurosis. The authors attacked their critics for their narrow misconception of the psychoanalytic view of sexuality. In cases of external danger they maintained the frustration was to self-love; that is, the injury was to the individual's narcissism. Ernest Jones, in a paper on that subject published with the others, pointed to the traumatic stress victims' lack of sociability, their lack of affection for friends and relatives, their feelings that they had been neglected or slighted and their importance insufficiently recognized as evidence of "wounded self-love."[9]

Freud, in the preface he wrote prior to the 1920 publication of the papers presented in Budapest, saw traumatic neurosis falling into the general category of narcissistic neuroses, but did not join Ferenczi, Abraham, and Simmel in asserting that observations of war neurosis confirmed the sexual basis of the disorder. Instead he stated that the matter remained to be resolved.

Freud had begun to approach the problem in a somewhat different direction, stimulated by another contradiction that traumatic neurosis seemed to pose to his theories. How were repetitive nightmares of combat to be reconciled with his belief that wish fulfillment was the motivating force behind dreams? Peacetime nightmares could be linked to punishment for forbidden sexual wishes, but repetitive combat nightmares could not be so easily explained.

Consideration of this question led Freud to a new view of trauma. In *Beyond the Pleasure Principle*,[10] he considered trauma to involve a breaking through of the individual's defense against stimuli *(reitzschutz)*. He saw traumatic neurosis as the result of fright *(schreck)* —a condition occurring when one encountered a danger without being adequately prepared. The repetitive dream was an attempt to be prepared after the fact, to dissipate by repetition the anxiety generated by the experience. For a comparable example of mastery by repetition, Freud used observations he had made of an infant who dealt with the anxiety of being separated from his mother by repeating a game in which he threw his toys away and then took considerable pleasure in finding them again.

Abram Kardiner's work on the subject, first published in 1941,[11] remains the basis for much of contemporary thinking on posttraumatic stress. He acknowledged Freud's *reitzschutz* theory as the starting point for his own thinking, which incorporated traumatic stress into an adaptational frame of reference. Kardiner saw trauma as an alteration in the individual's usual environment in which the adaptive maneuvers suitable to previous situations no longer sufficed. With the balance between the organism and its adaptive equipment broken, a new adaptation was not possible, and the individual accommodated his shrunken inner resources with the development of symptoms.

The constant features of traumatic neurosis that Kardiner saw were fixation on the trauma, repetitive nightmares, irritability, exaggerated reactions to unexpected noise (startle reactions), proclivity to explosive aggressive behavior, and a contraction of the general level of functioning, including intellectual ability. He also saw loss of interest in activity as a result of the breakup of organized channels of action which were replaced by periodic outbursts of disorganized aggression. The internal conception of the self became altered, confidence was lost, the world was seen as a hostile place, and the patient lived in perpetual dread of being overwhelmed. To Kardiner, combat nightmares were the result of this altered conception of oneself and the outer world.

During World War II, psychiatric efforts to understand traumatic stress were overshadowed by the immediate need to keep men functioning as combat soldiers. Psychiatrists were placed in combat areas, close ties between the soldier and his unit were emphasized, and the soldier's quick return to the combat group was considered essential.[12] "Morale" became a watchword, to be achieved by periods of rest and relaxation. In keeping with this goal, the symptoms of traumatic neurosis were subsumed under such terms as "combat fatigue" or "combat exhaustion," which were broad enough to include virtually every psychiatric disturbance seen among combat soldiers.[13]

The wartime Special Commission of Civilian Psychiatrists concluded that acute combat reactions were characterized by a "temporary psychological disorganization," with accompanying symptoms of irritability, insomnia, problems in memory and concentration, and flattening of affect. With the exception of recurrent nightmares, which were not described in the acute cases, the symptoms were quite similar to those previously identified among World War I veterans with chronic cases of war neurosis. The Commission was impressed with the fact that in 95 percent of the cases the acute reactions subsided after a brief period away from combat. The individual could then be returned to action.[14] Although there is no record of any attempts to see what happened to such individuals subsequently, the numbers of veterans with traumatic combat reactions seen in the decades after the war suggested the problems were far from temporary.

After World War II observations of concentration camp survivors, studies of the Japanese survivors of atomic explosions, studies of the survivors of natural disasters, and work with combat veterans added new perspectives to our understanding of posttraumatic stress.[15] Among all the groups, it became evident that in large numbers of cases the incapacitating consequences of trauma not only persisted, but often worsened when reenforced by stresses in the individual's subsequent life. Moreover, among both combat veterans and concentration camp survivors, many individuals were observed who seemed to adjust well immediately after their traumatic experience but later developed symptoms, in some cases many years afterward.[16] The concept of "delayed stress," so dramatically apparent in Vietnam combat veterans, was actually a reaffirmation of what had been observed in the earlier studies.

During this period, work with both soldiers and civilians exposed to life-threatening trauma began to emphasize the relationship of two contrasting aspects of the stress response: the tendency to reexperience the trauma in nightmares or daytime thoughts, images, and emotions; and a mechanism of numbing or emotional withdrawal. The repetition continued to be seen as a reflection of an effort toward completion or mastery of the trauma, while the numbing came to be considered an attempt to ward off or diminish the intensity of intolerable ideas and emotions.

Work with concentration camp survivors made clear in a way that past studies had not that guilt over surviving, when relatives or friends were killed, could play an important part in the stress disorder.[17] Sometimes the feelings of guilt were related to what the individual might have done to save others from dying, but more often such feelings were rooted in having survived. Robert Lifton, who studied Japanese survivors of Hiroshima, concluded that survivor guilt was at the heart of the posttraumatic stress reaction.[18] Subsequently, these reactions have been observed, although less ubiquitously, among combat veterans with posttraumatic stress disorders.

Care is required in making generalizations about

posttraumatic stress based on the study of any one particular population as there are different forms of expression for varying stress situations.[19] The fact that the
likelihood of survival for Jewish inmates of concentration camps was remote, and that those who died were
often close family members, undoubtedly contributed to
the high incidence of survivor guilt in this population.
A similar high casualty rate among family members
was probably responsible for the high frequency of survivor guilt among Japanese survivors of the atomic explosions of World War II.

Combat soldiers, however, vary in their emotional
relationships to their comrades. Our work suggests
that survivor guilt is more likely to be present in veterans who survived situations in which others who were
close to them were killed. On the other hand, traumatic stress reactions without survivor guilt are often
seen in combat soldiers as well as in civilians exposed
to life-threatening danger in which no one's life was
actually lost.

The absence of physical aggression in concentration
camp survivors has been documented, particularly
among those who reacted to their persecution with
overwhelming fear and a profound sense of helplessness.[20] But physical aggression, often explosive in character, is a common feature of the stress disorder in veterans of war. The soldier learned during combat to use
his aggression toward the enemy to reduce his fear,
epitomized in the basic act of firing his gun. That the
veteran who develops posttraumatic stress continues to
use his aggression to diminish fear in peacetime does
not seem so surprising. Physical aggression was not
compatible with survival in concentration camp inmates. Comparably, veterans who had traumatic combat experiences but never fired a weapon are a minority

whose posttraumatic stress disorders do not include explosive expressions of anger.

For many Vietnam veterans guilt, not over survival, but over combat behavior is of primary importance. Even when necessitated by combat, the killing of women, children, and the elderly often led to a posttraumatic stress reaction centered on guilt. A veteran who killed many enemy soldiers during his tour of duty painfully recalls only one killing, that of a woman armed with two grenades who came toward his squad. Today he wonders how he would have felt had his wife been killed by another man's act of violence.

The nature of combat in the Vietnam War left many soldiers unsure whether they were killing enemies or innocent bystanders. There was a rampant feeling that the whole country was the enemy and, therefore, the distinction between combatants and civilians did not much matter. Traditional moral reference points were blurred, leading many soldiers to react to life-threatening situations with diffuse aggression, often out of control—a reaction which led to a different kind of guilt.[21] In over one-third of the combat veterans we have evaluated, profound guilt over intentional actions of a nonmilitary nature—rape, mutilation of enemy dead, and killing of enemy prisoners or of other American soldiers —is at the core of their posttraumatic stress disorders. Clinical accounts of veterans in treatment after earlier wars do not suggest the presence of such actions with any comparable frequency. In chapter II we will discuss our study of combat veterans who have not developed posttraumatic stress. It is significant that none of the veterans in that group was involved in non-military violence.

Our work,[22] and that of others,[23] indicates that among Vietnam combat soldiers, who the individual was before exposure to life-threatening situations and how he behaved in them often play an important role in whether or not he develops posttraumatic stress. Such factors also play a decisive role in the form the disorder takes. These observations run counter to conventional wisdom since World War II, which has tended to minimize, if not altogether dismiss, the personal dimension as a significant determinant of posttraumatic stress disorders.

Character is more apt to play a role in posttraumatic stress disorders developed subsequent to combat than it does in most other forms of trauma, as we will see in the next chapter. A flood survivor or a Japanese survivor of Hiroshima was the victim of forces over which he or she had no control and survived, at least in part, by accident. Much of the death—and survival—in war is similarly accidental, but much of it is not. The soldier is often an active participant in the stress that may eventually overwhelm him.

When trauma is repetitive or sustained over time, as it is in war or was for concentration camp inmates, it is easier to see the role of individual character differences. Different types of disturbances have been noted, for example, among Nazi persecution survivors who experienced their ordeal in a passive way as opposed to those who were able to remain active and carry on some type of resistance.[24]

Similarly, veterans who functioned effectively in combat have developed stress symptoms with different features than those who did not. A soldier discussed in the next chapter, whose refusal to accept responsibility throughout combat contributed to the deaths of two men in his command, developed a posttraumatic stress disorder centering on the guilt he felt about this incident. This man's combat behavior was an extension of a prewar history marked by constant fighting with his parents over their demands that he assume more responsibility, initially for his younger siblings, and later for himself.

Neither the variations in the effects of different kinds of trauma nor differing individual responses to them was the focus of the American Psychiatric Association's task force that worked during the late 1970s to define criteria for the diagnosis of a posttraumatic stress disorder. Rather, the task force drew on work with combat veterans from World War I to Vietnam, with concentration camp survivors, and with victims of various civilian disasters to come up with symptoms that could serve a common diagnostic purpose.

As later defined in the Association's *Diagnostic and Statistical Manual of Mental Disorders* (DSM III),[25] the diagnosis depends on the presence of a "clearly identifiable traumatic stressor." Although traumatic stressor is not specifically defined, it is generally agreed that it involves an encounter with the possibility of sudden violent death. Memories and images associated with death are so central that Lifton speaks of such trauma as causing a "death imprint."[26] In this sense trauma means something quite different than it did in Freud's early use of the term to describe those psychologically conflicted situations in childhood that gave rise to symptoms. Further, traumatic stress should not be confused with the accumulated stresses of life, ranging from job pressure to divorce, that are referred to in social-psychological literature.[27]

In the presence of a clearly identifiable traumatic stressor, intrusive recollections of the experience—such as nightmares, unwelcome thoughts, or reliving experiences (flashbacks)—were deemed essential to the diagnosis, as was emotional numbing. The diagnostic criteria included six other symptoms, at least two of which had to be present to make the diagnosis: exaggerated startle response, insomnia, exacerbation of symptoms by events that recall the trauma, avoidance of such events, guilt over what might have been done to survive or guilt over surviving when others had died, and difficulty with memory and concentration. The first three of these symptoms are seen as manifestations of a continuing hyperalertness to potential danger, although in the case of insomnia, the danger feared is usually nightmares. The difficulty with concentration and memory is usually regarded as part of the general ego contraction described by Abram Kardiner. It is also a reflection of the individual's self-absorption with the distracting ongoing effects of the traumatic experience.

Anxiety, depression, and irritability with explosive outbursts of anger were listed as frequently associated symptoms, but not as necessary criteria for the diagnosis of posttraumatic stress disorder. The fact that these criteria do not include explosive outbursts of anger as an essential symptom reflects the need to include only symptoms seen commonly in all those with posttraumatic stress, that is, the need to include combat veterans and concentration camp survivors under the same umbrella.

DSM III represents an essentially successful attempt to embrace within one diagnostic framework the consequences of very different traumatic experiences. These consequences have enough in common to make the criteria eminently usable for diagnostic purposes, but that should not blur important differences among survivors of different kinds of traumatic stress.

Knowledge of either the specific stress symptoms produced or the nature of the trauma that produced them is not sufficient to understand or treat posttraumatic stress disorders.[28] Our own and others' work with combat veterans[29] has made clear that the response to the trauma of war is not only a function of the events which the individual encountered, but also of the specific ways he experienced those events, what they meant to him while in combat, and what they continue to represent for him in the postcombat period.

3

The Meanings
of Combat

COMBAT may be intrinsically traumatic, but is it traumatic in the same way for everyone who has been exposed to it? Many have explored the role of the individual's pre-existing character and personality in determining the development of the disorder following exposure to life-threatening trauma. The majority of psychoanalysts who observed stress disorders in response to combat in the years immediately following World War II agreed with earlier analysts concerning the central role of childhood precursors.

Leon Saul considered that infantile and childhood conflicts centering on hostility played a central role in the development and persistence of stress symptoms among veterans exposed to combat in World War II.[1] These symptoms, especially the belligerence often observed in the stress reaction, were thought to be a structured and purposeful expression of neurotic conflicts that had previously been repressed. Theodore Lidz similarly reported finding a history of faulty family relationships, with residual hostility and exaggerated sensitivity to the withdrawal of affection, among World War II veterans with severe, persistent postwar nightmares of their combat experiences.[2] He saw these nightmares as conflicted expressions of suicidal wishes in which death was both sought to end anxiety, and feared because it was the final separation from sources of affection.

Abram Kardiner was the first psychoanalyst to take a sharply different point of view.[3] He indicated that, unlike the pathology in hysteria, a traumatic experience in which one's survival is threatened can touch off a reaction that need not be of symbolic significance. Though realizing that a preexisting neurosis or psychosis could be exacerbated by wartime experiences, he warned against assuming an established predisposition as the basis for symptom development, and concluded that the basic psychopathology of the traumatic neurosis was not rooted in earlier conflicts. Subsequent clinical work confirmed the view that, with sufficient stress, anyone will develop a "war neurosis."[4]

Work since World War II with a variety of traumatic stress victims has tended to support the thesis that personality has little impact on whether traumatic stress develops. Leo Eitinger's study of 226 concentration camp survivors in Norway showed that 99 percent had some psychiatric disturbance years afterward. Eighty-seven percent had persistent nervousness and irritability, 60 percent had sleep disturbances, and 52 percent had continuing nightmares of their experiences. The majority of these cases could be diagnosed as suffering from posttraumatic stress.[5]

Most observers of civilian trauma victims have likewise concluded that if the stress is great enough almost everyone exposed to it will develop strikingly similar stress responses.[6] Frederick Hocking, an Australian who reviewed a large number of published accounts of the consequences of traumatic stress, found, "There was no correlation between the symptoms and the preexisting personality or any other factors in the patient's earlier life." Assuming exposure to stress of a fixed duration and degree, Hocking concluded that preexisting personality characteristics do little more than determine how long an individual can tolerate the situation before the onset of neurotic symptoms.[7]

The belief that factors preceding trauma are essentially irrelevant to the stress response has a natural corollary in finding the meaning of the trauma directly in the experience itself. Since we know that in war those who are exposed to sustained combat are most likely to develop stress problems afterward, it is not surprising that work on posttraumatic stress disorders in Vietnam veterans has focused primarily on the war experience itself. Much of this work, either directly or indirectly, emphasizes the ubiquitous quality of the combat stimulus and the veteran's response to it in a way that omits the personal and subjective quality of the experience.

Chaim Shatan[8] and Sarah Haley[9] called attention to stress reactions triggered by guilt arising from participating in or witnessing inhumane actions in combat. Lifton saw reactions precipitated by the very act of surviving a life-threatening situation in which others have died as the paramount problem with which Vietnam veterans are struggling.[10] John Wilson, in an attempt to develop a conceptual schema, classified combat stress on a scale of ego development based on the work of Erik Erikson, viewing events such as the death of combat buddies as activating identity conflicts. Guerilla warfare, Wilson suggested, arouses anxiety about trust from an earlier state of psychosocial development.[11]

Although these analyses have alerted us to some of the psychodynamic mechanisms linking combat experiences to the postcombat stress response, our work with a large number of Vietnam combat veterans—both those who have developed posttraumatic stress disorders and those who have not—finds very little support for the implication that a particular traumatic event has an intrinsic meaning that determines the individual's response to it. War itself is inhumane, yet actions that profoundly disturb one man do not so deeply trouble another, at least not in the same way. Guilt at surviving when others have died, for example, is far from universal, even among veterans with posttraumatic stress disorders.

Among veterans exposed to a considerable amount of combat, we have found that the development of a stress reaction is as integrally related to how the individual experienced specific combat events as it is to such objective dimensions as the content, intensity, or duration of the events per se. Several of the Vietnam veterans we have seen have witnessed or participated in virtually identical combat actions, yet the same event had quite different meanings for each soldier.

We use the term "meanings of combat" to refer to the veteran's subjective, often unconscious, perception of the traumatic events of combat. Our experience has been that these meanings usually derive from a complex meshing of factors related both to the individual's prewar life and to what he encountered in Vietnam.[12]

The Meanings of Combat

This concept permits us to attempt to resolve some of the confusion surrounding the role of preexisting personality factors in the development of the postcombat stress disorder. Those who consider such factors to be critical see the stress syndrome as a more usual form of neurosis. Those who see them as having minimal importance in the development of posttraumatic stress disorders view the traumatic event as sufficiently and inherently stressful to result in long-term reactions.

This dichotomy is conceptually misleading. The fact that a posttraumatic stress disorder may develop in anyone subjected to a sufficient amount of stress does not mean that precombat personality factors do not play a critical role. Who the veteran was before combat is highly significant, and sometimes decisive, in shaping the way in which the posttraumatic stress disorder develops. This does not imply that a personality disorder was present in the veteran or would have developed without the stress of combat, or even that such individuals are more vulnerable to stress in any general sense. Rather, it is indicative of the importance of close examination of precombat, as well as combat factors, in attempting to understand the meaning of the war experience to the individual and his postwar adaptation to it. Following are brief case reports of three veterans who were each evaluated by us during the course of five one-hour interviews. The cases illustrate that how the soldier experienced specific combat events can be as important as the events themselves in determining the nature of the posttraumatic stress responses.

Ted Ford was a powerfully built thirty-year-old veteran with straight black hair and a thick mustache. His appearance and gruff manner were in sharp contrast to the insecure and dependent picture that emerged from his story. When we saw him, he was separated from his wife and had been involved for a year-and-a-half with a woman he was planning to marry. He was working as a maintenance employee in a factory, having lost several previous jobs because of a drinking problem that originated during his combat tour in Vietnam.

Ted was the third of four children who grew up in a working-class family in upstate New York. From the time he was quite small, Ted had felt jealous of the attention his mother gave his younger brother. His sister and older brother, he felt, were favored by his father; his older brother, however, was protective of Ted, and their relationship was close.

Ted disliked his father's strictness and got into frequent arguments at home. In the tenth grade he dropped out of school because he did not like being told what to do. He worked for several years in a grocery store, which he hated, and when the draft threatened wanted to go to Canada. His father, a World War II combat veteran, insisted that he had to serve. Ted enlisted, but to deny feeling coerced by his father, exploited his enlistment to show his friends he was a big shot.

Ted was homesick in basic training, intimidated by the drill instructors, and resentful of the hard time he felt they gave him. In Vietnam he was assigned to transportation and, in addition to driving a truck, had a considerable amount of patrol and bunker duty.

During his first exposure to combat in Vietnam, Ted was terrified and refused to leave his truck to go on patrol. The first time his outfit was shelled Ted screamed for his mother, dove head-first into a ditch, and sustained a mild concussion. A more experienced and somewhat older squad member took Ted under his wing. Ted subsequently performed adequately, was promoted to corporal, and received a medal for meritorious service in combat.

While he was overseas Ted began to drink to ease his tension. Some months after returning home he developed a posttraumatic stress disorder and became a problem drinker. Twelve years later he was still having recurrent thoughts and nightmares about his first combat experience when he refused to leave his truck, and also about a prisoner he saw pushed out of a helicopter. Another frequent nightmare concerned a baby he had killed when, under night attack, he had fired straight ahead without looking above the trench he was in. He hit an armed Vietcong woman whose baby was strapped to her back.

Ted's recollections were integrated around the common theme of vulnerability—the helplessness of the baby and the prisoner, and his own sense of being a helpless, frightened prisoner of circumstances beyond his control. When Ted discussed the war, the most frequent theme was his envy of those who were wounded; he envied their pensions and the attention they received because of their disabilities. In other areas of his life, Ted's recurrent, interrelated themes were his feeling that he never got a fair deal, his sensitivity to slights or a lack of attention on the part of his mother and girlfriend, and his lifelong envy of his mother's favored treatment of his younger brother.

A pleasant dream Ted had had several times since his return from Vietnam linked his family experience with the service. It concerned an actual visit that Bob Hope and singer Connie Stevens had made to his base. Before Connie Stevens began to sing a song titled "Just My Bill," she asked for someone named Bill to come up on stage. Ted volunteered, explaining that he had a brother named Bill, but the singer replied that she wanted someone whose own name was Bill. In the dream, however, Ted was chosen. Connie Stevens sang to him, the show was televised, and his mother saw it back home. In the dream Ted received the attention and recognition that he felt had been denied him during childhood, the service, and his current life.

Ted had married and had a child a few years after his return from Vietnam, but his heavy drinking, erratic working, and desire to have his wife be exclusively attentive to his needs led to her divorcing him after several years of marriage. He insisted that to attain extra care and attention he would have willingly been wounded or crippled in combat.

Ted saw himself as having been wounded and crippled psychologically early in life by his parents' indifference; combat reinforced and cemented his image of himself as injured and helpless. Combat also gave him a sense of being entitled to demand from his parents, his former wife, and his current girlfriend the interest and attention he felt had been lacking in his life. He continued to struggle with his wish to use his damaged self-image to avoid responsibilities, working regularly as a maintenance man but constantly hoping to find some way to avoid having to work at all.

The cases of Ted Ford and of other Vietnam veterans we have seen suggest that the response to combat is not simply a function of what occurred, but depends on how events and actions were experienced by the individual. The second veteran had an experience similar to that of Ted Ford's. He too witnessed a prisoner pushed to his death from a helicopter, but his reaction to it was remarkably different.

Don Gray was a tall, dark thirty-two-year-old veteran who worked in a print shop. He had been married for three years and, when we initially saw him, was experiencing some difficulty communicating with his wife, particularly about his experiences in Vietnam. This was often an issue in his relationships with others, as well; however, he had a number of good friends and an active social life.

Don had grown up in a suburb outside a large metropolitan area. When he was four years old his mother had a "nervous breakdown" and was hospitalized for over a year. Both his parents paid little attention to their children; Don and his older brother and sister raised themselves, supporting one another as best they could. There was little communication among them, with each one working and going his or her own way. Don did well in high school and after graduation took a job as a printer's apprentice. He was happy in his work when, at the age of twenty-one, he was drafted.

Don was a model soldier, rated highest in his basic training outfit. He was assigned to an air mobile unit in Vietnam that was sent on search-and-destroy missions in the jungle for several months at a time. On these missions Don was exposed to every imaginable combat stress, had a confirmed "kill count" of 15, and received many decorations and medals. He dealt with combat with hypervigilance and a protective reaction toward the men in his squad. He was proud that none of his men was killed while he was in Vietnam.

During the year after he returned home Don developed a posttraumatic stress disorder, spent his time drinking, and did not work until his money ran out. Intrusive, disturbing thoughts of combat centered first on a Vietcong soldier he had taken as a prisoner. The prisoner was transferred to the charge of other troops who took him up in a helicopter for interrogation. He was pushed out of it, and Don, who was on the ground below, saw him fall. Because he had captured this prisoner, Don felt some responsibility for his death, and believed that he could have prevented the incident had he been in the helicopter.

Don was even more disturbed by an event he heard about after he had returned home. The inexperienced soldier who replaced Don as squad leader ordered a squad member into a well in a search for Vietcong firearms, and the man was killed by a booby trap. Don knew he would have thrown a grenade into the well instead of ordering it searched. It was this event that triggered his posttraumatic stress reaction, centered on the belief that, had he remained in Vietnam as squad leader, this soldier would not have died.

Don had learned self-sufficiency and protectiveness toward others from his early family experience. Having learned to deal with his feelings by himself, he found it difficult to express what he felt in his work or in his marriage. His resentment of enforced early self-sufficiency was reflected in his irritated preoccupation with irresponsible co-workers.

Although he had experienced overpowering frustration of his early needs for care and attention, he managed a successful adaptation as a boy by becoming self-sufficient and responsible, and not sharing his feelings with others. While he was in Vietnam, Don's sense of responsibility widened into the protective role he adopted toward his men, permitting him to repress the fears generated by combat. Yet combat also threatened this lifelong adaptation, making him vulnerable in a distinct way to the deaths of the squad member and the prisoner.

The meaning of combat for Don was further illuminated by the adaptation that he continued in civilian life; for instance, joining a local veterans organization to help raise money for better services for Vietnam veterans. It is significant that Don's becoming a father, shortly before he was first seen by us, served to reestablish him in a protective role, and coincided with a marked lessening of his stress symptoms and the end of his excessive drinking. His need to hide his emotions, however, and the irritation he felt toward others who were irresponsible remained.

The service and combat histories of Ted Ford and Don Gray prior to the traumatic events that were responsible for their stress reactions are reflections of differences in their characters. Ted Ford's fearful reaction during basic training and his first exposure to combat were consistent with his general behavior before, during, and after the war. Don Gray, on the other hand, took his entire service experience, beginning with basic training, as a challenge to be mastered. Just as Ted's ability to arouse a sympathetic protective reaction in a more experienced older person had helped him to survive, so too did Don's vigilance and protectiveness of others. Both of their adaptations to combat and their posttraumatic stress reactions reflected the strengths and weaknesses each man brought to the war.

The third veteran, Dave Templeton, was suffering from the consequences of his avoidance of responsibility while in Vietnam. His behavior in combat colored his posttraumatic stress reaction, but that behavior was strongly linked to his precombat adaptation.

Dave was a stocky, bearded thirty-year-old veteran who had scars on his neck, the result of injuries he had received in Vietnam. He was divorced, living with a much younger woman, and working part-time in the post office. His evaluation began while he was participating in an inpatient alcohol treatment program.

Dave was the son of a small town sheriff and had been raised in an atmosphere of strict and rigid discipline.

His childhood and adolescence were marked by frequent fights with his father, who drank heavily, as did his mother. Dave perceived responsibility as an act of submission and resisted helping care for his much younger siblings. He also felt guilty about the abuse his mother had suffered defending him from his father's excessive demands.

During high school, Dave excelled in athletics but did poorly academically. He dropped out in eleventh grade and joined the Army in order to learn a trade and find some direction. He shared his father's positive sentiments toward the Vietnam War and hoped his enlistment would earn his father's approval.

In basic training Dave learned mechanics and was promoted to the rank of corporal. He was assigned to Vietnam as part of an armored vehicle squad and within several months was promoted to sergeant. In his ninth month in Vietnam, Dave's squad was ordered to return alone from enemy territory to the base. Although the order violated the standard practice of returning with at least one other squad, Dave proceeded as instructed.

Dave and his men ran over a mine and their vehicle exploded. Dave was seriously wounded but fortunately was blown out of the turret in which he had been riding. The two men in his command were trapped in the flames. Unable to help them, he witnessed the agony of their deaths.

Dave believed that he should have been driving, although it was not normally his responsibility. Soon after this incident, he developed a posttraumatic stress disorder involving repetitive nightmares of the explosion and a strong sense of guilt over what had happened. His guilt related to the fact that, as a sergeant, he had never accepted responsibility for his men, trying instead to be one of them. Not challenging a dangerous and foolish order was part of his refusal to accept responsibility. He linked being spared at someone else's expense with not having been "in the driver's seat," that is, with not responsibly assuming control. At the same time, Dave became enraged with the Army, feeling it was responsible for the situation in which he found himself.

After Vietnam, Dave's problems intensified, his stress symptoms became chronic, and he became a heavy drinker. He failed to carry out his responsibilities at numerous jobs and was unable to function responsibly as a husband or father. After five years of marriage, his wife divorced him.

During the time Dave was being seen he asked to borrow his girlfriend's car. She agreed, after exacting a promise that he would be careful with it. That night Dave dreamed that he had wrecked the car. In the dream he tried to excuse himself to his girlfriend by blaming another driver who had cut him off.

As in the dream, Dave's resentment of being asked to be responsible, his impulse to destroy the object of his responsibility, and his subsequent guilt had plagued him in reality. Dave perceived combat, as well as much of his early life, as a source of anxiety because of his inability to accept responsibility while resenting those who expected it of him. These factors contributed to his subsequent postcombat stress disorder, which centered on guilt.

The meaning of combat for Dave was shaped by his early sense of himself as irresponsible, a sense that was intensified manyfold by his combat experiences, and continued to influence his postwar life. Through combat, his irresponsibility had become linked with responsibility for the death of others. The reluctance to accept responsibility that this soldier brought to combat would likely have given him trouble as an adult whether or not he had gone to Vietnam. The impact of this reluctance on his combat experiences and on his subsequent stress reaction, however, seriously complicated his postwar life.

In contrast to Dave Templeton, Don Gray, who functioned well in combat, and who both before and during the war demonstrated a capacity for caring, supportive relationships, probably would not have developed difficulties without combat. Yet his preexisting personality was an integral part of the meaning he gave to his combat experiences, and directly affected the form and expression of his posttraumatic stress disorder.

Identifying the unique meanings of combat for individual veterans permits us to distinguish and understand the differing reactions of Don Gray and Ted Ford to witnessing enemy prisoners being pushed from helicopters. Both perceived the event as traumatic. For Ted it aroused an identification with the helplessness of the prisoner, and for Don a sense that he could have prevented what occurred.

The emotional state that accompanied combat experiences and often continues to accompany postcombat recollections plays an important role in shaping the meanings of combat. Although the most obvious emotion is fear linked to the threat of survival, we see in Vietnam veterans a number of additional emotional states which are associated with their combat experiences. Many veterans, such as Ted Ford, express overwhelming fear in response to the threat of annihilation. For some, rage becomes so dominant a means of dealing with the threat of death that it threatens the individual in a different, but comparable way, with loss of control. For others, such as Dave Templeton, guilt over behavior in combat is predominant. For Don Gray, his tense hypervigilance was followed by self-criticism, and a sense of not having lived up to the high expectations he had for himself.

Not only are there varied emotions associated with combat, but even where the same emotional response is present, the subjective experience may be quite different for different veterans, reflecting the unique self that the individual brought to the war. The guilt that troubled Dave Templeton derived from the traumatic consequences of his longstanding refusal to accept responsibility. Failure to accept responsibility is but one of the many different sources of guilt we explore in chapter 5.

It is important to distinguish the meanings of combat from conscious reconstructions or interpretations of combat events. Among concentration camp survivors,[13] Japanese survivors of the atomic bombings,[14] and survivors of such disasters as the flood in Buffalo Creek, West Virginia,[15] the inability of those involved to find a satisfactory explanation for what happened has been described as contributing to their difficulties in resolving the experience. The difficulty these survivors have had in finding an adequate explanation has social as well as personal dimensions; the individual's sense of continuity within the fabric of his or her community was largely destroyed in each of these stressful experiences.

Vietnam veterans certainly share this difficulty. Their sense of a war fought without the total commitment of the government or the full support of the American people, their lack of clear understanding of the military or political objectives of the war, and their inability at times simply to identify the enemy, all helped persuade them that there was something casual and unnecessary about the killing and dying in Vietnam. Their attempt to come to terms with the personal meanings of their combat experiences is confounded by nam. Their attempt to come to terms with the personal meanings of their combat experiences is confounded by their socially shared sense of suspicion and dislocation from a country that to many seemed indifferent about their lives.

For many who fought, the American government's failure to pursue the war to a successful conclusion heightened their feelings that their sacrifices were meaningless. Even in combat veterans who do not have a posttraumatic stress disorder, there generally is residual bitterness. Some of those with the disorder link their difficulties to a sense of failure over not having been permitted to stay in Vietnam until the war was won. But in these veterans this sentiment is usually compounded by a deeper sense of personal failure related to what their combat experiences meant and still mean to them. Veterans suffering from stress symptoms are often more readily in touch with their social and political alienation than they are with the personal ways that Vietnam touched them.

In the process of understanding these personal meanings of combat we can begin to comprehend the full complexity of how this war has impacted on the men who participated in it. Although in the course of a comprehensive diagnostic evaluation[16] it is usually possible to make an initial assessment of the meanings of an individual's combat experiences, it is in psychotherapy that such meanings can be fully determined and elaborated.

Treatment

4

Stress-Oriented Psychotherapy

SOLDIERS respond to the trauma of combat in highly personal ways, and aspects of experiences that continue to trouble them in the postwar years vary considerably. Often what troubles them cannot be inferred even from the knowledge of what the individual did or what he encountered in Vietnam. Yet it is the unresolved nature of these experiences, which are often recalled in an incomplete, distorted form, or not at all, that keeps veterans tied to the war more than a decade after coming home.

The essential goal of psychotherapeutic treatment of the combat-derived stress disorder is to open up and explore in detail the content of specific traumatic combat experiences with the veteran, his feelings about these experiences both at the time of their occurrence and currently, and the particular meanings which the experiences had and continue to have for him in light of both his precombat and postcombat life. We have found that only through direct, sustained, and repeated explorations of combat events, feelings, and meanings can veterans achieve sufficient resolution of their experiences to make Vietnam no longer dominant in their postwar behavior and emotions.

In using the term "stress-oriented psychotherapy" we wish to emphasize the essential focus of treatment on the stress disorder and its roots—perceptual as well as objective—in specific combat experiences. All too frequently, we have encountered veterans who have spent years in treatment dealing with work or marital problems, or drug or alcohol abuse without discussing the combat experiences and posttraumatic stress disorder that were the source of these difficulties.

From the time the psychological symptoms caused by trauma were first recognized, treatment of the disorder was seen as a unique psychotherapeutic problem. Initially, the symptoms were seen by Breuer and Freud as amenable to the "talking cure" (catharsis), in which the patient was helped to recover repressed traumatic memories and to express the emotions associated with them (abreaction).[1] When Freud realized that catharsis and abreaction were not usually sufficient to effect a cure, emphasis shifted to childhood conflicts, which were relived and worked out in the relationship with the therapist.[2] Later a consensus emerged that the development of posttraumatic stress is not dependent on childhood precursors, and the roles of both abreaction and catharsis came to be reconsidered in the treatment of reactions to trauma.

The role given to catharsis by most modern observers is somewhat different from that attributed to it by

Freud. For Abram Kardiner, the critical therapeutic factor was not merely the expression of traumatic memories and the emotions associated with these, but also involved systematic reconstruction of the traumatic event and filling in of the amnesia that accompanied trauma.[3] Kardiner documented the successful use of short-term psychotherapy in the treatment of two World War I veterans seen six or seven years after their combat experience. In each of the cases, single traumatic combat experiences were responsible for a variety of symptoms and inhibitions. The patients could not recall many of the details of the experiences, but with the aid of their dreams Kardiner was able to reconstruct with them what had happened. These events were gone over many times until the patient could "tolerate the recollection of these details with no untoward effects." Then the more difficult task began of showing the patient the relation of his current inhibitions to the traumatic event.

In working with World War II soldiers with acute war neurosis, Roy Grinker and his associates[4] produced emotional recall of traumatic combat events through the use of "narcoanalysis," a procedure in which patients were interviewed under the influence of barbiturates.[5] Although in some cases positive results were achieved in a short period of time, these clinicians eventually concluded that additional therapeutic sessions without medication were necessary to allow a full working through of the traumatic material. Although narcoanalysis and hypnosis are sometimes still used to diagnose posttraumatic stress or to stimulate recall of traumatic events,[6] our experience is consistent with the conclusion reached by others who have worked with Vietnam veterans that the recovery of memories and the emotions associated with them regularly occurs in the context of effective psychotherapy, and generally does not need to be elicited through drugs or hypnosis.[7]

Many patients with posttraumatic stress, moreover, do recall the details of the traumatic events that precipitated their stress disorder, and may utilize a variety of defenses other than repression and amnesia. A thorough description of the defensive maneuvers utilized by posttraumatic stress patients is contained in the writings of Mardi Horowitz.[8] He points out that some patients will recall a traumatic event in words but will not picture the experience. Others will recall traumatic events but will not relate them to anything else in their lives; the events remain isolated and unintegrated. Still others will recall traumatic events in detail, but will express no feeling in connection with the recall. Horowitz emphasized the role of therapy in helping patients to achieve an integration of the cognitive and emotional elements in their response to trauma.

Most of Horowitz's therapeutic attention is given to civilian cases seen relatively soon after the trauma, an opportune time for exclusive focus on reactions to it, especially since acute symptoms are usually responsible for bringing the patient to seek help. Although in our experience with Vietnam veterans we see cases in which posttraumatic stress disorders have become florid more than a decade after the combat experience that induced them, usually we are dealing with individuals who have had the disorder for ten to fifteen years, twice as long as the most chronic cases described by Kardiner. Are treatment approaches derived from past therapeutic experience applicable to the treatment of posttraumatic stress in Vietnam veterans seen so many years after the trauma? Our work suggests that in some cases they are, while in others different approaches are necessary.

Stress-Oriented Psychotherapy

We have found that some veterans have managed, albeit at a great price, to separate their stress disorders from the rest of their lives. These men can work and function adequately in their families apart from episodes of stress symptomatology. With others, every aspect of their civilian life is affected by their efforts to cope with their stress symptoms. With these veterans the responses to combat trauma have often become merged with the reactions to subsequent stressful events such as the loss of employment, separation from family, or problems with the law. These stressful civilian events must also become the focus of the therapeutic intervention. Here, principles we have learned from therapy with posttraumatic stress victims must be integrated with principles more generally applicable to short-term psychotherapy if treatment is to be effective.

Another key difference between most veterans we have seen and other stressed populations whose treatment has been reported concerns the nature of the trauma to which they were exposed. Most case descriptions concern individuals who are reacting to a single traumatic event: an accident, a disaster, or an instance of sudden loss of a loved one. The therapeutic task in such instances, if not simpler, is more focused. With combat veterans several specific traumatic experiences usually converge in the disorder. Often a variety of traumatic combat events experienced by the veteran will be organized around certain themes unique to his particular perception. Understanding the common meanings of such experiences is central to the treatment of the many veterans for whom the disorder is not confined to one overwhelming traumatic event.[9]

In order for the therapist to discern what combat has meant to the individual veteran, a thorough knowledge of specific traumatic events must first be obtained. As

Horowitz has noted, because individuals have a variety of ways to defend themselves against trauma, the therapist may have to make a number of different interventions during the process of attempting to explore combat experiences. In some cases, veterans may try to avoid any discussion of combat, focusing instead on current difficulties. Especially when the individual is undergoing a crisis, current circumstances will need therapeutic attention; when a posttraumatic stress disorder is present, however, the source of current difficulties in unresolved combat experiences must be suspected, and this will necessitate opening up the issue of combat at the earliest opportunity.

Even in cases where veterans are willing or even eager to talk about combat, specific details of the experience can be obscured or events can be related with no sense of the individual's involvement, reaction, or emotion. Here, too, while respecting the veteran's need to set his own manageable pace, the therapist must continually pursue the essential task of reconstructing combat trauma until the entire setting, events, behaviors, and accompanying feelings are communicated and understood. Even though it is important to recognize that additional details of traumatic events which add to the therapist's understanding frequently emerge much later in the therapeutic process, the initial process of exploration is not complete until the therapist has a sense of comprehending the essence of the particular event or action.

In attempting to identify what specific combat events meant and continue to mean to veterans, we have found that combat dreams and nightmares provide invaluable clues. They are also of critical help in treating the disorder.[10] In the context of individual stress-oriented psychotherapy veterans should be encouraged to relate their dreams and nightmares and to associate details of the trauma connected to them. Repressed details and feelings connected with specific combat events often emerge only after the event has been repeatedly discussed over an extended period, but specific parts of dreams can often point to aspects of the veteran's experience that have not emerged through conscious reconstruction. The particular detail may seem minor, but if it has not been related before there is usually significant emotion bound up in it. Moreover, the events in civilian life that precipitate combat nightmares help to clarify the ongoing ways in which trauma distorts current perceptions.

As veterans recover from posttraumatic stress disorders their dreams reflect their progress. Jack Edwards, a veteran discussed in chapter 6, recurrently dreamed that he was helpless and immobilized during a combat assault. After he had made considerable progress in resolving his stress disorder he dreamed that he was triumphant in the same type of assault. Similarly, veterans who have coped relatively successfully with traumatic combat events often have dreams in which they are anticipating the trauma that actually befell them but in the dream either the worst part of the experience does not take place or they deal more successfully with the trauma. Michael Drake, discussed in chapter 7, dreamed of an ambush in which his platoon was almost wiped out and he was wounded. In the dream, however, he is waiting for the attack, is in a better position to shoot back at the enemy than he was in reality, and is not wounded.

Another essential aid in identifying and understanding combat events is the veteran's flashbacks, revisualizations, or reliving experiences. Although some who have worked with Holocaust survivors have reported that treatment possibilities are limited among those who relive their traumatic experiences in a state of altered consciousness, [11] we have not found this to be true for combat veterans.[12] Systematically exploring the content of reliving experiences provides a valuable therapeutic tool for understanding the meaning of the veteran's combat experiences and for eliciting the accompanying emotions. A cessation of this form of reexperiencing is also an important sign of therapeutic progress.

Warren Saunders, a former helicopter pilot discussed in chapter 8, frequently relived Vietnam combat experiences. Reliving an incident in which he had been shot, he would fall backward. While driving he would relive a time in which his helicopter went out of control and he would actually drive off the road. As the fear related to these combat events diminished in his therapy, he moved from reacting to the events, to perceiving them as unreal, to recognizing that they were happening only in his head. In time the anxiety and palpitations accompanying his recollections ceased, and though they remained unpleasant memories, they were no longer disruptive.

Although knowledge of the veteran's specific combat experiences underlies our understanding of his posttraumatic stress reaction, in our experience an understanding of precombat factors is often an important part of stress-oriented psychotherapy. In this respect, our approach differs from that of Kardiner and Horowitz who both saw the posttraumatic stress reaction as essentially rooted in and determined by the particular traumatic experience. Neither explores the role of preexisting personality factors in determining the particular way in which the trauma was perceived and experienced or the way in which the stress disorder is manifested. We have found, however, that understanding the veteran's precombat personality and functioning facilitates an assessment of his adaptive resources prior to combat which influenced how combat events were experienced.

A comprehensive understanding of both precombat and combat factors, as these converge in the meanings the veteran gives to specific combat events, is usually an essential element in the successful treatment of posttraumatic stress disorder. For Dave Templeton, a veteran discussed in the previous chapter, successful resolution of the guilt and subsequent need for self-punishment that comprised his reaction to the combat loss of his men necessitated exploration of his long-standing resentment against having to be responsible for others.

Having a general knowledge of the veteran's precombat functioning also helps the therapist to understand how a posttraumatic stress disorder is expressed in the context of the individual's personality and character. Virtually all veterans with posttraumatic stress, for example, show a markedly diminished capacity for pleasure (anhedonia), supporting Henry Krystal's observation that this condition is a "characteristic and reliable indication of posttraumatic states."[13] We have found that with progress in therapy, however, as fear, anger, and guilt cease to dominate the veteran's life, a remarkable improvement in the capacity for pleasure takes place, at least among those veterans who had demonstrated an ability to enjoy themselves and their relationships with others prior to service. In the case of veterans whose preservice lives indicate an impaired pleasure capacity, some degree of anhedonia is likely to remain despite the resolution of the basic posttraumatic stress disorder.

Finally, with regard to the significance of precombat experience, it may be noted that Kardiner assumed that attempts by veterans to link the traumatic events of combat with earlier life experiences represented retrospective attempts to defend themselves against the stress disorder. Our experience with Vietnam veterans is that they seldom try to make such linkages. If anything, they tend to see their lives as beginning—and often ending—with combat. The ability to see combat in the context of other life experiences is usually a sign of a lessening of the stress disorder.

Particular attention in the treatment of Vietnam veterans must be given to the relationship between the therapist and the veteran. Although all combat veterans may be expected to feel some initial reluctance or mistrust when attempting to communicate their experiences to a therapist, with veterans of the Vietnam War this tendency often appears to be especially pronounced because of the lack of support and understanding they received from those at home, during and following their time in Vietnam. Obviously there needs to be a minimum core of trust for therapy to begin, but many veterans, we have found, make the primary determinant of whether such trust is possible the manner in which the therapist deals with their combat experiences.

Supervising the treatment of Vietnam veterans, we have frequently found that therapy flounders at the point when the veteran has shared at least some of the specifics of his combat tour. The therapist may respond to stories of cruelty or brutality in combat with revulsion, or fear that such behavior indicates a potential for violence and possibly personal danger. Some veterans will try to intimidate the therapist in this way, but usually cease these tactics if they see the therapist is not frightened or angered.

More frequently, the therapist's discomfort is communicated in the need to convey understanding and acceptance before he or she is really in a position to do so. This reaction often reflects an unconscious desire to curtail unpleasant communications related to combat. In other cases, the therapist may not feel personally threatened by the veteran's potential aggression, but may be afraid that elaboration of combat descriptions will interfere with his or her ability to like and respect the veteran. When this happens, it is the therapist's discomfort rather than what is specifically said that the veteran responds to, leaving him feeling even more unacceptable and isolated.

Veterans who felt their aggression was out of control in combat often see any relationship, including the therapeutic one, as having the potential to release their destructiveness. Before exploration and relief of some of the distress produced by traumatic experiences, however, most veterans are not able to explore in any significant way the complexity of emotions that will develop toward a therapist. Usually only after the therapist has helped the veteran to begin to deal with the continuing impact of combat on his postwar relationships with family, friends, and co-workers will the veteran permit his feelings to be explored directly in relation to the therapist. The dreams of veterans with posttraumatic stress provide an early clue that combat trauma rather than feelings toward the therapist should be dealt with first. In the repetitive combat nightmares of most veterans, the therapist does not usually appear, even in disguised form. In our experience, it is only after considerable therapeutic progress has been made that combat-derived nightmares become less stereotyped and the therapist may appear in them.

Another salient factor in stress-oriented psychotherapy with Vietnam veterans is the difficulty most have in managing the powerful feelings that are set off by discussion of their wartime experiences. In our work we have found the intensity of emotions associated with recalling and reconstructing traumatic combat events dictates that the veteran be seen no more than twice a week, and in many cases, only once a week. Frequently, veterans will "forget" or consciously choose to miss appointments when the sessions become particularly painful. We have found it important to respect these attempts to regulate the therapeutic pace. In this respect, our experience is consistent with the observations of Elizabeth Zetzel, who treated British soldiers in World War II,[14] and Henry Krystal, who treated survivors of the Holocaust,[15] that therapy must be tuned to the "affect tolerance" of the traumatized individual. As we have indicated, however, the affect tolerance of the therapist is of comparable significance in determining the outcome in treatment.

If during the treatment symptoms such as anxiety, depression, rage, or insomnia become too incapacitating, a limited amount of medication may provide necessary but temporary relief. Medication may be particularly necessary for a subgroup of veterans, described by Lawrence Kolb,[16] who appear to have a conditioned emotional response to combat-related stimuli at the core of their stress disorders.

Although a considerable amount has been written about chemotherapeutic treatment of posttraumatic stress, the specific objectives and effects of such treatment remain largely unknown due to the lack of systematic, large-scale study.[17] There is ample support, however, for our conclusion that behavioral rather than pharmacologic interventions are preferred whenever possible.[18] Specifically, since the veteran's dreams and nightmares provide the fundamental insight into what continues to trouble him about his war experiences, medications which aim at their suppression should be avoided whenever possible.

Out of the Vietnam experience other techniques have been used in the attempt to help veterans work through the disturbing effects of combat. Perhaps the most widely adopted of these has been the use of group therapy, which derived largely from the early experiences of veterans' "rap groups."[19] Clinicians have reported beneficial outcomes among veterans, resulting from the mutual trust, understanding, and support supplied by the group.[20]

While acknowledging the value of the therapy group, our experience supports the conclusion of others that without the simultaneous process of individual psychotherapy (not necessarily with the same therapist), this technique in most cases is inadequate for veterans to work through their posttraumatic stress disorders.[21] Although the group can help the veteran feel less isolated through the sharing of common experiences and emotions, we have found that most veterans do not explore the most intensely personal meanings of combat trauma in a group setting.

Over the course of our work with Vietnam veterans we have consistently been impressed with the effectiveness of individual psychotherapy specifically focused on the trauma of combat. Although meanings of combat are largely personal, we have identified several different clusters of meanings commonly seen among Vietnam combat veterans that are manifested in sharply different forms of posttraumatic stress disorder. For some veterans, guilt over actions they performed or failed to perform in Vietnam pervades their combat recollections, and even when the guilt is unconsciously experienced, it is frequently reflected in self-destructive postwar behavior. Others are dominated by depression stemming from the combat loss of friends; in the case of still others, postwar life appears to be an extension of combat, marked by the same hypervigilance, anger, and proneness to attack developed in Vietnam. Each of these clusters of meanings presents particular therapeutic problems that will be explored in the following chapters.

5

Meanings of Guilt

GUILT, as we have seen, is often part of the stress response to trauma. In any traumatic situation in which people feel frightened or helpless, there is a deep-rooted tendency to blame oneself, that is, to treat one's fate as a punishment that in some sense must be deserved.[1] The propensity for guilt in traumatic situations is sometimes a more direct outgrowth of fear: for example, when a soldier runs in terror from combat and torments himself with self-reproach afterward.

Guilt during combat is usually seen as a reaction to killing. When killing is not necessary or warranted, or is committed in a mood of angry retaliation, or especially when it is committed in a non-combat situation, susceptibility to guilt is even greater. After World War II, guilt over having killed prisoners of war was a factor in the stress disorders of some American veterans.[2] In those Vietnam veterans whose killings involved women, children, and the elderly, guilt is apt to be an integral part of the stress disorder.[3]

In some cases, combat-related guilt appears to the outside observer to have an objective and understandable basis. This was the case with Dave Templeton, the veteran discussed in chapter 3, whose reluctance to accept responsibility contributed both to the trauma that befell him and the deaths of the crew of his armored vehicle. More often, however, the situation in combat is not nearly so clearcut. With veteran Dennis Allen, guilt was more a subjective function of who he was, rather than what he did or did not do in combat. Only in the most vague and diffuse sense was he in touch with the experiences that triggered his guilt; he was not at all in touch with the meaning they had for him.

Dennis did not develop the symptoms of post-traumatic stress until 1981, twelve years after he left Vietnam. During the evaluation interviews that preceded his therapy, he presented a series of disparate, painful Vietnam memories whose common meanings were not discernible. He was thirty-six-years-old and had been a physical therapist for several years. His stress symptoms had developed after his job began to bring him in contact with disabled Vietnam veterans who sometimes spoke of their combat experiences. He had been married for seven years and was the father of two children. He saw his marriage as being without significant problems, although he had been somewhat removed from his family since his difficulties had begun.

From August 1968 to August 1969, Dennis had served as a medic with an American advisory team assigned to the South Vietnamese Army. In the first half of his tour he was stationed at a large base where he treated both American and South Vietnamese soldiers, as well as

South Vietnamese civilian casualties. During this period, the base came under mortar attacks and shellings. During the following six months, Dennis also experienced frequent sniper fire when he traveled by motor vehicle or helicopter to treat the wounded at various field bases. Since he did not have to go into heavy combat zones, however, he considered his assignment relatively safe.

Dennis felt he had only been under moderate stress in Vietnam, most of which involved experiences with sick, wounded, dying, or dead soldiers and civilians. He considered himself to have been proficient in his work, was pleased he had had an assignment in which he had saved lives, and felt his self-confidence and ability to be assertive had increased during the war.

In describing specific experiences he had in Vietnam, he recalled in a pained, constricted manner an American soldier who had been evacuated from the field with half his face shot off and who eventually died. He also recalled several disturbing incidents that involved children. In one he was criticized by a nurse for failing to give CPR (cardio-pulmonary resuscitation) to a little girl whom he had been sure was dead. In another, he had separated a baby and a mother during a helicopter evacuation, and since no records were kept, he later realized it was likely that the child had become an orphan. A final incident involved the refusal of a soldier who was tending bar at the base to give him ice for a feverish child. Dennis came back with a gun and got the ice.

He also spoke of a young American serviceman, assigned as a sniper on riverboat patrols, with whom he became friendly. The first time the young man shot someone he became excited and went on shore to look at his victim. The wounded enemy soldier exploded a grenade, blowing himself up, killing a South Vietnamese soldier who was nearby, and seriously wounding Dennis' friend. Dennis felt that he had probably died, since his wounds were severe, but he did not know for certain since the young man was evacuated out of the area. In describing the incident, Dennis said, "It was stupid of the kid. You don't walk up to someone you've just shot once and don't know whether he's alive or dead." Relating this story, his voice broke frequently.

The night following his departure from Vietnam Dennis' base suffered its first direct attack, and there were several casualties. He felt lucky that he missed the attack, but was troubled by the thought that he could have been of help.

After the service Dennis finished college, completed training as a physical therapist, married, and had children. Except for an episode in 1975, when he became upset after seeing a character in a movie with half his head shot off, he thought the war was behind him. Increasingly, however, he found himself preoccupied by memories of his own experiences, and felt more and more withdrawn from people. He also felt guilty over having suffered so much less than the disabled veterans with whom he was working.

Initially, Dennis presented his Vietnam experiences as a series of discrete events, describing them cursorily, with little emotion other than tension and without elaborating their significance or meaning for him. He would try instead to engage in intellectual discussions about posttraumatic stress disorders. Although these defenses were challenged, it was not until he was able to arrange his work so that he was not exclusively involved with Vietnam veterans, that his anxiety was sufficiently reduced to allow him to begin to explore his own combat experiences and their meaning for him. He was seen in individual therapy once a week over a three-month period.

At first Dennis was eager to come for his sessions, though they aroused his anxiety. Early in his treatment, he related a dream. He was on an archaeological dig and it was exciting, but there was also some fear and danger associated with it, and there were helicopters overhead. The people he was with found a skeleton. The helicopters landed and took some of the people out. He was left behind, but he felt the helicopters would come back.

The mixture of eagerness and anxiety were what Dennis had felt while talking about his life in his sessions. He was able to see the archaeological dig as a metaphor for the searching into his life that we were doing. The fact that looking into his past would bring up skeletons and would be associated with death became more evident in subsequent sessions.

Talking further about the American soldier who had half his face blown off, Dennis recalled that he had helped take this man out of the helicopter. He remembered feeling angry at another member of the company who was taking pictures of the wounded soldier. He felt that no one had any business taking "an interesting picture" at such a time. In response to a question about how he felt looking at the wounded soldier, he said the sight was "grotesque." One eye was gone and half the face, but he was driven to look and felt odd about the impulse to do so. Recounting this, he suddenly realized that he too had taken a "picture" of what had happened.

Dennis had watched the doctors attempt to save the soldier. Watching the surgery he felt sick, which had never happened to him before. To get away from the experience, he tried to busy himself, attending to two other soldiers who were not so seriously wounded.

Meanwhile, the doctors botched a tracheotomy and the soldier bled profusely and died. Dennis felt angry at the doctors but glad the soldier had died quickly, since it seemed clear he had little chance of surviving in any case.

He also felt anger toward his young sniper friend for not having been more self-protective. Since it was the impulse to look at the person he had killed that had cost his friend his life, the incident confirmed Dennis' sense that looking at death was dangerous, destructive, and potentially fatal.

He related this to his overall sense that he had been an onlooker in Vietnam, someone who observed the casualties of combat without being part of the action. Being a medic on a base, not in the field, contributed to this feeling. Being assigned to a South Vietnamese Army unit rather than to an American one made it even stronger. At the time he had been glad for the greater safety of his assignments, but subsequently he felt that he had missed out on something.

His anger with the man who took the picture of the wounded soldier was related to his sense of himself as someone who stood back from involvement. Realizing only afterward the full consequence of separating the mother and the baby was partly the result of his lack of empathic involvement. This lack also made him doubt whether or not he should have given CPR to the child he had been sure was dead. He liked to relate the incident about getting the ice for the sick child because it reassured him of his willingness to become involved. Yet the link between involvement and potential violence bothered him in that incident as well because he had used a gun to get it. He was torn between perceptions of detached curiosity as a hostile act, and involvement as a dangerous or violent one.

As Dennis was able to experience the feelings and associations related to his combat experiences, and as he became able to perceive their common underlying themes, he became less of an onlooker in his therapy and in his life. After an initial period of anxiety, his symptoms markedly abated. As they did, there was considerable improvement in his relationship with his family and his satisfaction at work.

The combat experiences of Tom Bradley, the next veteran, had more of the stamp of Vietnam in terms of their guilt-arousing potential. As his case illustrates, however, a strong, unwritten code of conduct encouraged many combat soldiers in Vietnam to deny or repress the guilt they felt over actions that were not strictly military in nature.

Although Tom's entire postcombat life had been pervaded by a posttraumatic stress disorder and he had already spent over a year in therapy, when initially seen by us he was not at all in touch with the guilt at the core of his disorder. For years Tom had managed to suppress most of his symptoms of posttraumatic stress with LSD and intravenously administered amphetamine, both of which he began to use shortly after his return from Vietnam. In the mid-1970s, when he stopped using the drugs because he felt they were destroying his body, he became aware of how preoccupied with Vietnam he was, how pervasive his combat nightmares were, how difficult he found it to be with people, and how easily he became enraged. He remained a heavy marijuana smoker, feeling he still needed marijuana to reduce the intensity of his stress symptoms and, particularly, the frequency of his nightmares.

Tom was a tall, thirty-four-year-old veteran with curly red hair and a full beard. He spoke in a serious manner and appeared both angry and depressed, yet he related well to people and had a fair range of emotion. He was married, had two children, and worked as a factory technician. His job was marked by frequent absences, and he was conscious of taking out the frustration and anger generated by the pressures at work on his family.

Tom had served in an infantry unit in Vietnam during 1967–68 and saw extensive combat, particularly in the Iron Triangle area north of Saigon. His most intense recollections concerned three specific incidents. One involved the first Claymore mine he set off, which killed an armed Vietcong woman. In the second incident his unit was ambushed by a North Vietnamese Army unit making its way out of Saigon. After throwing grenades at and killing two enemy soldiers who were firing on his squad, he went into a nearby hut. The hut had been booby-trapped and Tom was wounded and knocked unconscious. When he woke up his gun was gone and he assumed the enemy had taken it and left him for dead. He considered himself lucky that they had not put a bullet in his head to make sure. He spent a month recuperating from the effects of a concussion and wounds in both legs.

The third incident involved a mountain-top patrol near a Buddhist monastery. His squad was acting as security for engineers making maps when they were infiltrated by the enemy via underground tunnels and lost several men. The infantry unit that came to their aid set fire to the monastery and killed the priests in retaliation for failing to alert the Americans to the presence of the tunnels.

Tom reported two recurrent combat nightmares. In one dream he is in Vietnam walking through thick grass in the "point" or lead position. Suddenly a head gets in the way and he slices it off. In reality, he had walked point for a week, but nothing comparable had occurred. Upon questioning, he remembered that several times he and his squad had been instructed to dig up Vietcong graves and search them for hidden weapons. During these assignments, they had sometimes knocked off the heads of the corpses with their shovels.

The second nightmare, he said, "... scares the hell out of me. It's so real but I don't know if it actually happened." In the dream he is carrying the dead body of a young woman and trying to hide it so no one can find it. Upon waking from this dream he would sense that he had some involvement in the girl's death, but would be unable to recall what it was. In a third recurrent nightmare he is under attack on the monastery hilltop; with him are his wife and children.

Tom was seen ten times in a three-month period in therapy that focused on his combat experiences. We spoke first about the incident with the Claymore mine. Tom had gone over to see the body because he wanted to establish that it was his kill. He had some discomfort that it turned out to be a woman, but since she was armed and everyone seemed to be proud of him, he said he had felt some pride.

He said he had no feelings about looking at the woman's body and recalled, "Vietnamese bodies always seemed like mannequins. They never seemed real, like Americans." He added that they always bloated faster and bled less than the bodies of American soldiers.

In the incident in which his unit was ambushed he had a similar reaction. Describing the two enemy soldiers he killed, he said, "There was something surrealistic about the way the two gooks kept popping out of the trees. They reminded me of target dummies." He said he had tried to find the bodies because the "big macho thing was to have a kill count," but they had been dragged away.

No such protective dehumanization was possible when Tom encountered American dead. Once his company was sent to reinforce a base that had been overrun in what Tom called "a human wave attack" in which many American soldiers had been killed. Their bodies looked grotesque to him. He felt horror, shock, fear, and a sense that "it could have been me it happened to." He recalled that his eyes had started to tear, but noted, "You learned not to show such reactions."

In the course of his treatment Tom was encouraged to describe again the time his squad's position was penetrated while they were protecting the monastery. He revealed that the night before that event his squad had been visited by a group of Vietnamese prostitutes. The men had wondered later whether the prostitutes had revealed their position to the Vietcong infiltrators. On the night of the attack, Tom's unit was stationed in pairs around the mountain, spelling each other for sleep. In the morning two men were found in their cots, their throats cut. Presumably both had fallen asleep, perhaps because they had had less sleep the night before because of the prostitutes' visit.

Tom said everyone in the squad was certain they had been betrayed by the priests in the monastery. The men had been friendly with the priests, who must have known about the tunnels but had not warned them. Tom felt the experience had made him cynical and mistrustful.

Tom dissociated himself from what had happened when the support unit came in and took control of the mountain top. From his position on an armored personnel carrier waiting to go down the mountain, he saw the place go up in smoke, heard some shots, and could recall thinking it possible that everyone in the monastery had been wiped out. He seemed not to want to know for certain what had occurred, but indicated he could understand the anger that had triggered this retaliation and acted as though he was not bothered by it. This manner of defending himself against guilt became more evident as his experiences were probed further.

When asked if he had ever raped any Vietnamese women, Tom replied that he had not, but added casually that he had once witnessed a rape. He was asked to elaborate. He said that his squad had been assigned to secure the entrance to a tunnel complex while four men from another squad went underground to explore the tunnels. He was in radio contact with the other squad and learned that they had found an underground Vietcong hospital base. They reported finding a French doctor, two French nurses, and a Vietnamese assistant. A short while later Tom said he heard shouting and the sound of grenades exploding. The four men came out of the tunnel dragging one of the nurses, who was bleeding from arm wounds. Each of the four raped the nurse while Tom's squad watched. When the last man was finished he pulled out his knife and killed the woman. When this happened, Tom and his squad departed; he never knew how the men disposed of the nurse's body. He did know that when the four soldiers reported the incident they made no mention of taking anyone alive.

Tom claimed to have had no particular reaction to the event; he understood the anger of the four men, who had been in combat for a long time. When asked if he had been sexually excited watching what had happened, he freely admitted that he had been.

Although Tom had never before connected this episode with the nightmare in which he was trying to bury a dead girl, he was able to understand that it was the stimulus for the dream. He realized he had "participated" in the rape and in the cover-up of what had occurred. Just as in the dream he was trying to find a place to bury the dead girl's body, he had tried for years to bury the entire event. Although he had succeeded in this on a conscious level, the burden of guilt he was nonetheless carrying was evident in his dream.

Tom was in touch with the fear he had experienced in combat, but not with the intense feelings attached to hacking the heads off the corpses, or to his vicarious participation in the rape and murder of the nurse and the retaliation against the priests. The meaning of his stress symptomatology was rooted in his guilt in relation to these experiences. Of equal importance was his attempt to protect himself from the awareness that he was deeply troubled by what had occurred. It was necessary to challenge his efforts at denial and to make him aware of his involvement and his feelings of guilt. As this occurred, he began to feel less controlled by his combat experiences.

Tom's nightmare of being with his family while under attack in Vietnam suggested his perception that his combat experiences threatened his relations with them. There was ample confirmation of this in his behavior. By taking out his anger and frustrations at work on his family, he saw them as defenseless and comparable to those who were helpless in Vietnam, like the French nurse or the priests, who were also recipients of anger generated elsewhere.

Tom was aware even more critically that when he felt closest to his wife, he needed to pull away from her. With his children, pleasure in playing with them would be interrupted by surges of irritability. He connected his enjoyment of his family with a slackening of the perpetual vigilance he had learned in Vietnam, specifically with the disaster that resulted when he and his squad let down their guard at the mountain-top monastery. As his posttraumatic stress symptoms lessened Tom became more relaxed with his family, but it was uncertain at the time he ended treatment how deep or lasting these changes had been.

In contrast to Dennis Allen and Tom Bradley, Bill Clark was typical of those veterans who are acutely aware of their guilt over events in combat, and whose ability to repress that guilt ended when they left Vietnam. Bill was not so aware of the need for punishment that his guilt had produced and the degree to which this need had influenced his behavior since coming home from Vietnam.

Bill was a stocky, muscular man of thirty-five who was married, had four children, and had worked steadily as a mechanic for the last fourteen years. He had been in Vietnam in 1966, assigned to operate a machine gun on an armored personnel carrier (APC). He had served with various infantry units first in the Cu Chi area north of Saigon and then farther north around Pleiku. Soon after his return from Vietnam he married, but during the postwar years gradually withdrew from his family, spending more and more time alone in his bedroom, thinking and reading about Vietnam. In recent years he had also been drinking heavily.

Outstanding in his recollections of combat was the revulsion he had felt being exposed to dead and mutilated American soldiers, whose bodies he often transported back to the base in the APC. Even more disturbing was an incident in which a friend had been captured and killed by the Vietcong. The mutilated body, the genitals cut off and stuffed in the soldier's mouth, had been left by the Vietcong outside Bill's camp. Although everyone knew that the Vietcong's intention had been to frighten them, the officers insisted that the men look at the body in order to fuel their anger and combativeness. Bill recalled feeling sick at what he saw and regretted that he had not turned his eyes away as others had done.

He was also troubled by his awareness that by shooting in "free-fire zones" he and his outfit had killed unarmed civilians, and by his recollections of involvements with Vietnamese prostitutes in which he had been physically abusive. Although Bill related his combat experiences with a good deal of emotion, at times becoming angry and at other times tearful, he was vague on specifics, claiming not to remember the details.

Bill's posttraumatic stress disorder was characterized by reliving experiences, which he described as "weird things I do that I have no memory of." When we initially met him he had just been hospitalized for the first time after getting drunk at a friend's house, wrecking furniture, then running out and shooting at imaginary Vietcong in a nearby field. A few months earlier, also when intoxicated, he had set a fire in his kitchen, shouting that he was burning out the Vietcong. During such episodes he sometimes referred to himself as a "mass murderer." He was sorrowful and distraught discussing these episodes, saying he had a lovely family and feared he might do something to harm them.

Not all of Bill's reliving experiences took place when he had been drinking. Often his wife found him crawling around the house at night with his gun, as though he were back in combat. Once, out hunting, he felt another hunter was firing at him and was about to fire back when his brother got him to stop. Several times during sex he had referred to his wife as "mama-san."

During another reliving experience he had pointed a gun at her while speaking in a sexually derogatory manner.

When not reliving his Vietnam experiences, he avoided talking about them, particularly with his wife. Yet the reliving experiences appeared to be a way of confessing to her and to others what troubled him about his behavior in combat.

Although some of Bill's behavior would seem bizarre viewed out of the context of his Vietnam experiences, he was not clinically evaluated as psychotic nor did psychological tests reveal him to be so. His Rorschach responses were filled with references to explosions and weapons, but the overall integration of his responses was good.

Bill was seen twice a week during his four-month hospitalization and once a week as an outpatient for a year afterward. At the beginning, he tried to use his "bad memory" to avoid dealing with events in Vietnam with clarity or specificity. He had even less recollection of the reliving episodes in which he behaved as though he were back in combat.

Bill's inability to remember and his vagueness were gently but consistently challenged and he was encouraged to repeat in detail individual combat events. Gradually he was able to both picture and relate the details of his experiences while expressing what he felt about them. He also began to recall events he had not previously mentioned. It gradually became possible to explore with him the meanings of his combat experiences.

Virtually all of Bill's reliving experiences turned out to be elaborations of combat events which troubled him. For example, his wild shooting spree in the field related to a sense that he had had during his last months in Vietnam, of being out of control and firing at "anything that moved." His setting the fire in the kitchen was connected to an incident in which he had refused to crawl down an underground tunnel in pursuit of Vietcong and persuaded his sergeant to burn the enemy out instead. The incident in which he almost shot another hunter resembled an episode during a firefight when he had killed two Americans he had mistaken for Vietcong. The incident with his wife and the gun bore a direct parallel to Bill's behavior with a prostitute he beat unconscious with his gun (and may have killed) after he discovered her going through a friend's wallet.

Bill's sense of himself as a "mass murderer" derived from situations in which he and other members of his outfit returned fire from snipers who were shooting at them from villages. He was aware that for the most part those they hit were civilians, and probably not even Vietcong sympathizers. On one occasion, his outfit's return fire had been criticized by their lieutenant as excessive; Bill admitted in therapy that it had been reflexive and out of control.

It was not "mass murder," however, but a more personal killing that pained Bill the most. While returning fire from a village, he had seen people running, fired at three figures, and watched them fall. When his unit advanced, he saw that he had killed two children and a woman in a pink blouse, whom he took to be their mother. In recalling this event, he identified with the husband of the woman and imagined how he would feel if he were to experience such a loss. After what he had done, he wondered if he deserved to have a wife and children.

Bill's need for atonement or punishment was suggested by several dreams he had during the period of his treatment. In one dream he was in Vietnam receiving incoming mortar fire. He could see the shells and one of them had his name written on it. The dream recalled an actual situation in which an officer giving the wrong coordinates had called in mortar fire on Bill's unit, killing two men and wounding forty others. The dream was also linked to the time he had fired on and killed the American soldiers, believing them to be Vietcong. In a sense he was now calling in the fire on himself and making sure that he was hit.

Bill's constant state of overpowering fear in Vietnam had much to do with the violent behavior he described as out of control. This behavior, and its underlying anxiety, became more understandable after a nightmare which he characterized, in an attempt at humor, as his "atrocity story." Bill had dreamed that a combat buddy was kicking the heads of dead Vietnamese like soccer balls. In actuality, Bill had once shot off the heads of dead Vietnamese and his friend had kicked them, getting blood on his boot. He and his friend had laughed while this was happening. The dream, however, was terrifying for Bill, and permitted us to explore the fear of death that had been covered over by his efforts to be callous and indifferent.

Bill was encouraged to discuss his Vietnam experiences with his wife. After some delay he did, on a weekend pass from the hospital. Elated, he described the weekend with her as their closest since they had been married. The following weekend he brought another Vietnam veteran home with him, ignored his wife, and stayed up drinking and talking about Vietnam with his friend. Bill became upset, talked as if he were back in Vietnam, and smashed walls and broke glass in the house. The realization that he was breaking down the Vietnam wall that had separated him from his wife and family made him attempt to retreat behind it again. His new intimacy with his wife aroused the fear both of losing her and doing harm to her. After we explored this episode together, Bill stopped drinking and began to become closer to his family in a more sustained way.

When he was asked about sexual experiences in Vietnam at first Bill responded vaguely. But eventually he related, with detail and considerable feeling, several painful memories involving Vietnamese women. The first was an older prostitute whom he took into the woods and beat to unconsciousness when he was unable to perform sexually. The second was a younger prostitute who had once rejected him when he returned from the field looking bedraggled. Another time she agreed to go to bed with him, stimulated him to ejaculation but did not want to have intercourse. He became angry and raped her, beating her in the process. Another incident involved the prostitute whom he had left unconscious after beating her with his gun.

Bill connected his behavior with the women to receiving a "Dear John" letter from the girl he had been going with for several years before leaving for Vietnam. Telling his wife what had happened with the prostitutes and learning how much she had already guessed and accepted what had occurred was reassuring to him.

Bill's combat experiences were centered in violent, destructive behavior that he perceived as having been out of control. He was torn between a desire to hide such behavior and a desire to reveal it. Revelation through reliving experiences, in which he was in a state of altered consciousness, was a compromise resolution of this conflict. Bill felt he should be punished for what he had done in combat; both in his dreams and in his life he arranged his own punishment. He had wrecked an estimated fifteen cars, he had done much to wreck his marriage, and he had acted in a way that undermined his progress and promotions at work. How much his self-destructive behavior was the outgrowth of guilt over his Vietnam experiences was something of which he had little initial awareness.

As Bill discussed the various aspects of his combat experiences and their meaning for him, his memory became clearer, his ability to relate traumatic combat events with some specificity increased, and the power of these events to disturb him diminished. His combat nightmares became less frequent and he stopped having reliving experiences.

Bill left the hospital, but continued to come back as an outpatient on a once-a-week basis. He was glad to be home and was more involved with his wife and family than he had been for many years. He and his wife arranged to take a vacation together, something they had not done since their children were born.

As the time approached when Bill planned to return to work, however, he had a relapse. He became increasingly tense, his insomnia became severe, he spent his days hunting by himself, and his nights pacing through the house with a gun. He fantasized ambushing cars on the route to the hospital.

Now Bill expressed bitterness and disappointment with his therapy and came close to discontinuing. He could not cope with returning to work and dreaded the reaction of his co-workers to his hospitalization. He felt he would not be able to stand the pressure when he was busy at his job, though he had hated it when he was idle.

Shortly before Bill returned to work he had an extremely disturbing dream that was difficult for him to reveal. In it, Bill and another soldier set an ambush and killed some Vietcong. A woman and some children were in a ditch. The other soldier shot the woman and all except one of the children, then insisted that Bill shoot the remaining child. Instead Bill shot the other soldier in the head, but nothing happened. The soldier then shot Bill, also without effect. Then Bill's mother appeared with a gun, fired it into the face of his youngest son, and "blew him away."

The other soldier in the dream was a combat veteran whose abusive treatment of his wife and child Bill found disturbing. He himself had often had destructive fantasies of wiping out his entire family, particularly in periods when he felt especially vulnerable, a vulnerability he attempted to contradict by his imperviousness to the bullets. For the first time he confessed that before his hospitalization he had come close to suicide, an alternative he considered preferable to his destructiveness. But his daughter, who Bill had assumed was at school, discovered him with a gun pointed to his head. His shame at being seen by her put an end to his suicidal thoughts.

The boy who his mother killed in the dream was named after Bill. He was the child Bill considered the most vulnerable, and with whom he most identified. In the dream his mother seemed to represent all the people, including his wife and therapist, that Bill felt should be protecting him, but as he perceived it, were not. Although anger and the need for punishment were expressed in the dream, the image of his mother shooting her grandson reflected Bill's sense that no one could be counted on in a world gone mad.

In response to a question, Bill admitted that in recent weeks he had been having fantasies of killing the therapist. He related his reluctance to continue in therapy to this. He dreamed of coming to the therapist's office and shooting him in the head. He saw the therapist as having power over him through knowledge of his destructiveness.

Bill had also recently had a recurrent dream in which

he was trying to kill a black panther. Since the black panther had been the symbol of his combat unit, Bill saw the dream pointing to a struggle with something uncontrolled and violent within himself. In one variation of this dream, the therapist appeared in military fatigues to help him fight the panther. Then Bill noticed a game warden's insignia on the uniform the therapist was wearing. Discussing this, he said that he was apprehensive about the game warden because each hunting season he would kill more than the legal limit of deer. He was unsure whether he saw the therapist as someone who was helping him subdue his violence, or as the source of punishment for his destructive impulses. Dealing with these feelings was an obvious relief to him and marked a major turning point in his therapy.

Returning to work proved not to be as traumatic as Bill had anticipated. It became less disturbing for him to discuss his difficulties in trusting the therapist, his wife, and the men at work. He also started to talk more about his early life and future plans, an acknowledgment that he had previously had a life not dominated by Vietnam and might indeed have one again.

Although the three veterans discussed in this chapter all had posttraumatic stress disorders and were linked by certain common experiences in combat, they differed in how they perceived their experiences, and how they experienced and defended themselves against guilt. The three also reflect differences in how they manifested their posttraumatic stress, as well as the degree to which the symptomatology permeated their lives.

The guilt of one veteran over "looking" at death too closely, the denial of guilt of another who witnessed a rape and murder, and the guilt the third felt over his loss of control do not merely reflect different meanings of combat. Their varying reactions also represent different therapeutic problems. The fact that these men have a common disorder should not obscure the fact that understanding the unique features of each is essential to effective treatment.

Looking at traumatic events played a role, albeit a different one, in each of the veterans' adaptations to combat. For Dennis Allen, guilt over his curiosity in the face of death and the sense he had been an onlooker were central factors in his stress reaction. Tom Bradley tried to use his role as an onlooker to deny responsibility or guilt for the rape he witnessed. Awareness of his sexual excitement during the rape, and exploration of the dream in which he accepted responsibility for the death of a woman helped him to understand that he had been a much more active participant than he wished to acknowledge. Bill Clark, who was ordered to look at the mutilated body of his friend, did not feel guilt, but rather a sense that the same thing could happen to him. It was one of a series of frightening events that contributed to the loss of control of his aggression. That in turn led to actions that made him feel intense guilt.

Differences exist, as well, in the ways in which the veterans protected themselves against guilt. Dennis Allen and Tom Bradley repressed awareness that certain events had been traumatic for them. Dennis used the fact that he was helping the wounded soldier and his anger at another man for photographing him to avoid his own guilt for the curiosity that led him to take a mental picture of what happened. Tom used his belief that he was only a passive observer of the killing of the monks at the monastery and the rape and murder of the nurse to deny his guilt. His sense that he could understand the frustration and rage of soldiers who had participated more actively than he was a way of staying removed from personal responsibility. Taking the position that, "I did not do it, but can understand and not blame those who did," was Tom's way of avoiding the reality of what had occurred. His dreams were vital in opening up his defenses enough to permit him to come to terms with what had happened.

Recognition and acceptance of their guilt had immediate therapeutic value for Dennis Allen and Tom Bradley, but the problem was different for Bill Clark. He knew he felt guilty over the woman and children he had accidentally killed. He also felt guilt and shame over killing the two Americans he mistook for enemy soldiers. Most of all he was guilty about having been out of control so much of the time while in Vietnam. In his case, it was not his repression of guilt, but his need to confess and be punished for what he did that underlay his symptoms. His confessions in therapy and to his wife alleviated the need for his relieving experiences. The remaining therapeutic task was to relieve him of his need for self-destructive atonement. The fifteen-year sentence he had served in a prison of his own making was probably greater than any he would have received had he been punished for his combat behavior.

Reducing guilt reactions from traumatic combat experiences consistently reduces the residual fear attached to the experiences. Although guilt is often the outgrowth of fear, the development of guilt perpetuates and increases fear. The cycle of fear generating guilt which in turn increases fear, described in detail by Sandor Rado,[4] contributes to the intensification of post-traumatic stress disorders. The reciprocal relationship between guilt and fear, however, provides leverage for the therapist. When guilt is alleviated the cycle operates in reverse, to reduce fear and its expression in stress symptoms.

6

Combat Never Ends: The Paranoid Adaptation

A COMMON adaptation seen in Vietnam combat veterans with posttraumatic stress disorders can be described as "paranoid." This response to combat trauma involves eternal vigilance in dealings with others, an expectation that any argument is a prelude to a violent fight, and a need to strike first in the face of potential aggression. Under such emotional pressure, the veteran perceives civilian life as an extension of the war, and almost everyone not in his immediate circle of wife, girlfriend, family, or other combat veterans is seen as a potential enemy. Even with those who are important to them, these veterans usually find it difficult to show warmth. Many drink or take drugs in attempts to deal with their stress symptoms and inability to relax.

The weapons such veterans carry and their proclivity to attack anyone who inadvertently comes up behind them reflect a perpetual readiness for attack, even when no danger exists. More important, they display an underlying quality of total vulnerability. When confronted with overt hostility, the passage from vulnerability to counterattack is so rapid that the veteran may not be aware of the fear that underlies his response. His nightmares, which frequently do express his fear and vulnerability, can be triggered by currently perceived threats as well as memories of past dangers.

Features of the paranoid reaction may be found among most Vietnam veterans with posttraumatic stress since, as many observers have noted, hypervigilance, aggressiveness, and an accompanying denial of fear were so much a part of the combat experience.[1] The veterans we describe as having paranoid adaptations, however, meet the DSM III criteria for the diagnosis of a paranoid personality: mistrust, proneness to take offense, and restricted affectivity.[2] These paranoid qualities became evident only after they had developed a posttraumatic stress disorder following combat.

Other veterans with comparable paranoid features have borderline personalities, histories of psychopathic behavior, or both. Perhaps the most uncontrolled, destructive combat and postcombat behavior are witnessed in veterans with a combination of psychopathic and paranoid features. Such veterans usually have a precombat history of antisocial behavior. Although combat veterans who develop the paranoid syndrome we describe may show signs of belligerence, reflected in more than average involvement in fighting prior to service, they are not psychopathic; they exert more control over their emotions than do borderline or psychopathic personalities.

Past studies of postwar violent behavior in Vietnam veterans have related it to violence prior to combat and to "personal violence" in combat, that is, to actions that were not sanctioned by the military or that were directed against unarmed prisoners or civilians.[3] These studies usually do not identify violent combat veterans as psychotic, psychopathic, borderline, or paranoid personalities, nor do they ascertain whether or not their violent behavior occurs in the context of a posttraumatic stress disorder. Treatment considerations are often directed toward helping the individual find alternative ways of coping with anger, without reference to his overall psychological integration. This integration in our experience, is central to assessing both the veteran's potential for controlling violence, and his ability to gain insight into its sources.

Psychodynamic explanations of postcombat violence in veterans tend to be based upon conceptions of disturbed self-esteem that have been derived from the study of non-traumatic disorders.[4] In posttraumatic stress and the behavior that accompanies it, however, problems of self-esteem are secondary to a situation of emergency dyscontrol that is far from analogous to the sources of non-traumatic peacetime neurosis.[5]

About 20 percent of the more than a hundred Vietnam veterans with posttraumatic stress we have evaluated have a paranoid adaptation to the posttraumatic stress disorder. Frank O'Donnell, who was in therapy for a period of two years, is representative of this group. His case illustrates how combat experiences continue to shape the veteran's perceptions in civilian life and the task the therapist faces in ameliorating these perceptions.

Frank was a thirty-two-year-old veteran who had received a Silver Star, two Bronze Stars, and three Purple Hearts for service in Vietnam. He was handsome, well-built, well-dressed, and articulate, though somewhat tense and serious in manner. For the past few years he had worked as an administrator in a public school system. He was married and had two children, but was in the process of separating from his wife. Frank was referred to us by some veterans who suspected that he had war-related difficulties from the anxiety he exhibited when helicopters from a nearby military base flew over the school where he worked.

At the age of three, Frank was adopted from an orphanage by a couple who had already adopted another boy five years older than Frank. Since when his father became angry he threatened to send him away, Frank suspected he was adopted, but was not told the truth until he was thirteen. Although he expressed no anger toward them, he never felt close to his parents, nor did he feel they were close to each other.

Frank did well in high school, and was a sociable adolescent who liked to clown around. He got into frequent fights in bars, but compared the fights to picking up women—both were weekend forms of entertainment. Five weeks before he was drafted, he got married to a girl he had known most of his life. He said that he was more attracted to the closeness in her family, in contrast to his own, than he was in love with her.

Frank entered the service in 1969 and spent March 1970 to March 1971 in Vietnam, assigned as an infantryman to a cavalry unit. He was made point-man and is proud that, unlike most of the other point-men who were rotated, he always served in that position whether operating in a company or a platoon. Except for sixty days in May and June of 1970 when his outfit was in action in the Fish Hook area of Cambodia, he was in the Iron Triangle area of Vietnam. The two months in Cambodia, however, were the worst period for him, since his unit was involved in several firefights a day, as opposed to only one every few days in Vietnam. His most intense memories are of the Cambodian period.

In Cambodia, while looking for a suspected enemy base, Frank's company found a road cut into the jungle and set up ambushes. Frank was a forward observer who would signal the machine gunner when someone was coming down the trail. One time a North Vietnamese lieutenant came off the trail and walked straight toward him. "I had to kill him with my knife and the sucker wouldn't die. I must have stabbed him thirty times. I was kneeling on his chest and stabbing him and he was trying to get his pistol in my face. When I thought he was dead I dragged him behind an anthill. I was going back to my observation post when I saw the bushes start moving. The guy was trying to get up with blood all over his front, his neck, his face. So I had to go back over and kill him again. After I dealt with him a second time I went back to my position. We killed eight more soldiers [that day] coming down." Frank's unit stayed in the area two weeks, nicknaming the road "Redrock" because of the blood that covered it.

In early June, during an enemy ambush, Frank was wounded for the first time. His unit had been out in the field and had only five hundred yards to go to reach its firebase. It was the monsoon season and it was pouring. An inexperienced lieutenant in charge of their platoon insisted, against the advice of the men, that they spend the night in a bunker complex that they had cleaned out a few days earlier. The next morning, "I was sitting with a cup of Tang, which I had never had before and which a friend just gave me—then the whole world just blew up." From three sides, they were fired on by machine guns; the kill zone was right where Frank was sitting. The Tang flew out of his hand, his helmet was shot off his head, and his arm was knocked back by a bullet. Although he received only superficial wounds, three men within a few yards of him were killed, and seven were seriously wounded.

One week later, Frank was wounded again. On perimeter duty, he smelled dead fish and rice and discovered a small group of enemy soldiers. He called up his platoon and "we just blew that area away." The enemy counterattacked and he and another soldier were cut off. On the way back, "Maybe 15 yards ahead of me I saw this human hand come up out of the grass. It looked so out of place the first thing I did was shoot at it and the next thing I knew I saw it fly off. And then the next thing I knew we had fire coming from the front, both sides, everywhere. Trees above us were going down." A Chinese concussion grenade, which Frank assumed was thrown by the enemy soldier right before his hand was shot off, hit a branch of a tree, then hit Frank in the back without exploding. He and the other soldier jumped and rolled away, but the grenade blew up and wounded Frank in the leg.

The enemy retreated and on the way back to camp Frank walked up to the foxhole of the soldier whose hand he had shot off. "The hand was still there on the ground. I looked down, there's a North Vietnamese soldier looking up at me with a stump for a hand and I think he was getting ready to give up and the next thing

I knew I just unloaded the whole magazine into his face." Frank expressed regret that he had killed the soldier rather than allowing him to surrender.

His response to an incident that happened a short time later made him aware of his rage. On a clearing patrol, his platoon found two soldiers from their division hanging from a tree; the only thing left on their shirts was a cavalry patch. Both men had been shot many times and their genitals had been cut off and stuffed in their mouths. Frank said his outfit went a little crazy after that. They left cavalry patches on bodies and some of the men even carved the cavalry insignia into the skin of dead enemy soldiers.

Soon after returning to Vietnam from Cambodia, Frank came upon an enemy soldier who fired at him several times before he was able to shoot him. He went over to strip the body for "war trophies" and picked up the Soviet-made rifle the Vietnamese soldier had been carrying, which had a large bayonet. "The bayonet was out and for some strange reason I just took the whole rifle and just plunged it right in his chest as he was laying on the ground. I could never figure out why I did that."

By the end of the war Frank was in a mood to fight anyone from the enemy to his superiors. He received a disciplinary citation for punching a sergeant who had hidden during a firefight.

A month after Frank returned from Vietnam, he heard that the man assigned to replace him at point led the squad into an ambush; several men, including the point-man, were killed. Frank had liked this man and had helped to train him, but felt he was a poor soldier who "fell into more bunker holes than he found." Frank's first instance of postcombat violence happened the night he learned of this incident. A fight with his father-in-law and two brothers-in-law had to be stopped by the police.

When Frank first returned home, he was treated as a hero with a parade organized in his honor. He felt withdrawn and removed from people, however, particularly from his wife, perceiving that he had changed while she had remained the same. Frank considered her too dependent on her family and resented her preference for eating at her mother's home, which was in the same neighborhood. He felt she was not interested in sex with him and that she was generally disparaging and non-supportive. Whenever any difference occurred between them, he said, she would run to her mother. He did not accept any responsibility for the difficulties between them.

They had children, hoping to improve their relationship, but it only became worse. About five years after returning from Vietnam, he began seeing other women, and for several months before he was seen by us, he had been steadily dating one person. During this period he and his wife separated and reconciled several times. Finally Frank decided he wanted to end the marriage, although he felt nervous about it.

For eight years Frank was a policeman in his hometown. He believed he took the job because he "felt naked without a gun." When he quit, it was because he felt he was too quick to draw his gun in situations where it was not appropriate; one time, he shot at a group of youths who had broken into a soda machine.

After Vietnam, Frank had developed a serious drug problem with Darvon, initially prescribed as a pain-killer for his wounds. He continued to use the drug to offset his posttraumatic stress symptoms, in particular a debilitating anxiety and recurrent nightmares of his combat experiences. After a series of near fatal overdoses, he managed to get off Darvon, but he continued to be a regular marijuana smoker.

Despite Frank's heavy use of drugs and posttraumatic stress symptoms, while working as a policeman he managed to go to college at night, earn a B.A. degree, and begin graduate studies. He left the police force to work in the school system. At this time he became active in a veterans organization and spent most of his free time with other Vietnam veterans.

When we first saw him, Frank evidenced stress symptoms that had existed since his return from Vietnam more than a decade earlier. He frequently reexperienced combat in nightmares. Most painful was a recurrent dream in which he relived the incident in which he had repeatedly stabbed the North Vietnamese lieutenant. In his dream it seemed to happen in slow motion. Frank also had nightmares about the soldier whose hand he had shot off and whom he later killed, and about the man who was killed after replacing him at point.

In still another recurrent nightmare, he would dream that his helmet was being shot off and his arm knocked back, just as it happened when he was wounded and several men were killed. In the dream, however, some of the wounded and the dead would stand up and stare at him questioningly.

His nightmares contributed to severe insomnia, but he managed to work, often with only a few hours of sleep. His symptoms were exacerbated by events that recalled the war. Once, when his girlfriend offered him some Tang, his heart began to pound and he was gripped by anxiety.

The detailed, matter-of-fact style Frank used to describe his combat experiences and their traumatic manifestations in his current life at the beginning of therapy was a striking contrast to the powerful affect in his dreams about these experiences. When first challenged about this apparent detachment, he attributed it to his police training in giving unemotional reports. In time, and with further challenging, he was able to acknowledge the excitement and fear associated with the retelling of these experiences.

In most situations, Frank was extremely mild mannered. He tried to avoid getting angry, fearing his aggression would be out of control because of his tendency to see any physical encounter as a fight to the death. Sometimes direct involvement with situations that recalled combat led to explosive behavior. During a week in which he was intensely involved in meetings of the veterans organization and exposed to films of combat in Vietnam, Frank took a knife he used to carry as a policeman into a bar and began drinking. Someone insulted him and the next minute, "I had the fellow on the floor and was holding the knife to his throat."

Usually it was ordinary frustrations of civilian life that triggered excessive reactions and combat-derived nightmares. On a shopping trip a saleswoman was rude to Frank and he reported her to the manager. He said that had she been a man he would have jumped over the counter and beaten her up. That night he dreamed he was walking barefoot with his M-16 (as he had when he walked point in Vietnam), stalking a policeman who had mistreated him when he was on the force. He took the man out into the woods and emptied his rifle into him. By changing the incident which had occurred during the day into one involving a male adversary, he was more comfortable expressing his murderous feelings.

Other aspects of his civilian life also served as sources of "attacks" that Frank responded to as if back in combat. After he separated from his wife, she secretly appropriated his car, which was his prize possession. He had equipped it with a tape deck and a stereo and placed a decal of his Vietnam unit on the window. Twice that week he dreamed of the American soldiers he had found tied to a tree with their genitals stuffed into their mouths. In one instance the dream was followed by another about his killing of the Vietnamese soldier who had wounded him with a grenade. Frank perceived the loss of his car as an emasculation and reacted to it with anger appropriate to this perception.

At work, Frank usually controlled his anger, and did not carry a weapon. Once, however, he had difficulty with a co-worker who tried to block his promotion. Frank managed to circumvent the man's opposition and obtain the promotion, but his anger remained. He had fantasies of killing the man, and eventually provoked a fistfight with him that had to be broken up by co-workers. Following this incident, Frank began carrying his knife to work, claiming he was anticipating retaliation from the man's friends.

The problem was not that Frank did not know how to use his anger effectively in civilian life—in both the incident with the saleswoman and with his promotion, he was able to get what he wanted. At the same time, he saw physical action, in which he literally took the situation into his own hands, as providing the only effective relief. He perceived hostility from others in such magnified terms that murder was the only psychologically suitable revenge.

It was not easy for Frank to be aware of the fear and vulnerability underlying his rage. His recurrent dream of the North Vietnamese lieutenant he killed first enabled us to open up this subject. Frank revealed that after he had killed the man, he went through his wallet and saw pictures of his wife and family. He identified with him and realized that their situations could have been reversed. Consequently, he never again looked through the personal effects of dead enemy soldiers.

After some time in treatment, Frank began to reveal some of the most fearful aspects of his Vietnam experience. Once, when he was walking point, a baboon jumped from a tree directly in front of him and, even though Frank had a rifle and could have shot the animal, he felt paralyzed. He described the baboon as huge and black, with red eyes and an open mouth with fangs. The animal finally turned and walked away from him, but Frank could still recall the terror of the experience.

He also talked about the first time he had been in a firefight. His assignment was to supply ammunition to the squad's machine gunner, but when the call came for him to bring over more ammunition he froze, unable to move. Another soldier, noticing Frank's difficulty, took over the job. Although the other soldier never said anything to him about what had happened, Frank recalled that he had felt humiliated; he was unable to sleep and was anxious to redeem himself in combat. The other soldier's acceptance of Frank's fear led to a discussion with him about his intolerance of fear, his own, as well as others'. Picking a fight with the sergeant he felt had behaved in a cowardly manner was an outgrowth of this intolerance.

One incident helped Frank see the relationship between his continuing sense of vulnerability and his anger. Driving home after a pleasant day with his girlfriend, some young boys threw rocks at his car, and one hit the windshield. Thinking for a second it was a bullet, Frank reflexively ducked. Then he stopped the car, got out, and gave chase. He was barefoot, and even though the running was rough on his feet, he did not notice. He was high on the excitement of the chase and white hot with anger. He caught one of the culprits, a boy of ten who cried and pleaded not to be reported to his mother. Frank said that had it been somebody older he would have beaten him up; instead he let the frightened boy go. First exhilarated by this incident, he later became depressed.

That night Frank dreamed of the time when his outfit was ambushed, he was wounded, and several men were killed. The next time he was seen, he talked of his increasing interest in guns and how he would like to buy an M-1 carbine. He recognized that he equated the rock-throwing incident with the ambush in Vietnam. He saw too how he went from fear, to anger, to depression, comparing the last state to the low he always experienced in Vietnam after a "combat high." He was bound to have been depressed as well over the inappropriateness of his behavior.

Frank's combat nightmares expressed his depression as well as his fear and rage. During a period when he was financially unable to manage both his new apartment and the support payments to his children, his wife took him to court. She also refused to allow him to see the children. Frank reacted to her actions with resigned pessimism. He was doing nothing to improve the situation, saying he had nothing left to lose so he did not have to try.

On several occasions during this period, Frank dreamed he was walking point, as he had in Vietnam, except that he had tunnel vision—he was looking straight ahead without noticing what was on either side of him. In reality, walking that way would have been suicidal, since peripheral vision was essential to walking point effectively. The dream was a reflection of the narrow perspective he was taking in his to-hell-with-it-all attitude.

Another dream also recurred, with slight variations. In its most typical form, Frank was in a pond, wearing combat fatigues, and was being sucked down to its bottom. In one such dream, the pond was in back of his house, and he dove in the water because his younger daughter—the child he felt closest to—was being sucked down. He kept going farther and farther down, losing his breath. Finally, he was able to grab his daughter and pull her up.

Frank felt his security (and that of his daughter) threatened by the divorce. His emotions were the force that was out of control and were threatening him and the person he loved most. A central task in his therapy was to make him aware that the major threat to him now was internal.

When he was depressed, Frank's potential for violence was a source of comfort to him. He talked about his need for weapons and described how even in his girlfriend's town, where he knew no one, he carried a knife. In several dreams, he pictured having a grenade and experiencing a sense of well-being. Upon waking and realizing he did not have the grenade, he would feel sad.

He began to talk about the vulnerability he had felt since Vietnam, and his sense that attack might come from anywhere. He spoke in particular of how uncomfortable he was in a public place like a shopping center, always looking behind him and being fearful of people around him he did not know. He felt he protected himself by carrying a knife, but was aware of the danger inherent in his tendency to overreact with the weapon.

As much as Frank had been out of touch with the fear and vulnerability underlying his rage, he was equally out of touch with the guilt that surrounded his combat experiences. It was perhaps easiest for him to recognize the survivor guilt, reflected in the dream about the ambush in which three of his comrades were killed. He realized that the unspoken question in their rising from the dead and looking at him questioningly was, "How come you are alive and we are dead?"

Frank had expressed some regret about killing the enemy soldier he believed was trying to surrender. For the most part, however, he tried to justify what had occurred with a series of rationalizations: the soldier still had several grenades in his possession, he deserved to die because of the ambush his outfit was planning, and so on. It was some time before Frank could express the guilt he actually felt over that experience.

When initially asked about guilt, Frank related an incident that seemed at first trivial in comparison to other things he had revealed. Two Vietnamese children had gotten into a serious fight, and he and several of his friends gathered around to watch it. Afterward Frank felt that he should have broken the fight up. He was indicating what became clearer subsequently, guilt about taking pleasure in combat. In terms of his own experiences, he expressed this more directly, speaking of the excitement of combat and the "adrenaline high" it produced.

Frank had been proud of his effectiveness as point-man and liked being "where the action was." He missed the mixture of terror and excitement of landing by helicopter in a "hot LZ" (landing zone). His combat experiences, he explained, had been in "living color," while the rest of life was in "black and white." He once planned to add some excitement to a veterans' picnic by arriving in a helicopter with a few friends, but he gave up the idea when he realized that other veterans might find it upsetting.

After six months, Frank had made considerable improvement. His combat-related symptoms were less crippling, he had reached a better accommodation with his wife over the divorce, and he was doing well at work. Upset after seeing *Apocalypse Now,* he came home and cried openly in front of his girlfriend, confiding to her what it was in the movie that had upset him. This was the first instance, outside of therapy, that he had shared his feelings about combat with anyone.

Frank began to deal somewhat differently with other aspects of his Vietnam experience as well. When someone spoke critically of Vietnam veterans at a party, he left rather than get into a fight. He was also no longer so driven in his involvement with the veterans organization. His combat nightmares became less frequent and less intense; when he dreamed of the ambush in which his friends were killed, they no longer rose up to question his right to be alive.

But Frank still bore scars of his Vietnam experience. Physical scars from the wound on his leg caused him to avoid public swimming, less for cosmetic reasons than because of questions he anticipated. He continued to be uncomfortable in large groups of people, feeling he could not protect himself. He no longer, however, seemed to be in the grip of the terror and excitement associated with Vietnam.

Although Frank had considered his wife and children a necessary anchor during the stormy periods of his posttraumatic stress and drug abuse, he had shown little warmth or tenderness toward them. Following his separation from his wife, he accepted, without apparent disturbance, that she had turned his older daughter against him. How much of Frank's emotional unresponsiveness was part of his posttraumatic stress disorder, and how much it was part of the emotional coldness he had been brought up with, was not immediately clear.

During the early months of his treatment, Frank's passivity in dealings with his wife and their divorce was in marked opposition to the way he approached the rest of his life. Although he fantasized violence toward his wife, he felt he could not act on it because of the children. He ended up feeling paralyzed. As he came to understand and accept his feelings, the resignation that characterized his marital situation lessened. He reached a divorce settlement that included a financial agreement he was able to manage, and that gave him the right to see his children on a regular basis. The warmth and affection he could now demonstrate toward his girlfriend and children suggested that a significant part of his coldness had been the result of a mixture of his stress symptomatology together with his marital difficulties.

At this point Frank stopped coming to his sessions; he said he was feeling fine and the pressures of work were making it difficult for him to come in. When he finally did come in after several weeks, he related that recently, while falling asleep, he had experienced the feeling that he wanted to say something to someone, but was unable to make any sound. He said this was a feeling he had had before, and recalled a time in Vietnam when he had been in a foxhole and a mortar shell exploded nearby, causing a lot of earth to fall on top of him. He had felt as if he were being buried alive; he recalled screaming as he fought his way out.

Frank associated this cut-off feeling to the way he had isolated himself in the basement of his house when he was using drugs heavily. Looking out through the eye-level windows in the basement's concrete walls, he would feel like he was in a bunker. He admitted that sometimes he still had the impulse to cut himself off from everyone and was afraid of those feelings.

Frank then revealed his real reasons for avoiding therapy. Friends had told him that his therapist was responsible for problems that arose in the development of a treatment program for Vietnam veterans. He said he could not reconcile the disparaging things he had heard with his own favorable opinion. It had been easier for him to avoid the sessions rather than deal with his ambivalent emotions.

After talking about it, Frank was ready to deal not only with his ambivalent feelings toward his therapist, but with his need to avoid any friction in his deepening relationship with his girlfriend. His sense since combat that anger was such an all-encompassing emotion it was incompatible with affection and respect, made avoidance of it seem necessary to him. At the same time, he realized he would have to reconcile these contradictory emotions if he were to maintain any lasting close relationships.

The main features of Frank O'Donnell's story are similar to those we have observed in other Vietnam veterans whose adaptations to their posttraumatic stress disorders are of the paranoid type. The tendency to identify primarily with other combat veterans like themselves, to treat the outside world as the enemy, and to come alive in a climate of combat, is more than just a way of treating civilian life as an extension of combat. We have indicated that the meaning of combat experience to the veteran is an integral factor in the stress response. For veterans with a paranoid adaptation the meaning of the combat experience is linked to their angry response to their vulnerability. Anger serves to help them overcome fear and deny guilt.

Frank's combat adaptation, in which the rage he felt

seemed appropriate to the situation, but was not under control (as when he bayoneted a man he had already killed) is prototypical for veterans with a paranoid adaptation. Even in combat such behavior is questionably adaptive. Frank recognized that by shooting the wounded soldier who was trying to surrender he was taking a chance of giving away his position when the enemy was still around him.

Although when paranoid behavior continues in civilian life its non-adaptive function is more apparent, the veteran derives an illusory sense of comfort from the power associated with his potential destructiveness. Frank liked to carry a knife and derived pleasure from dreams in which he was holding a hand grenade and fantasies of killing those he did not like. Other veterans fantasize setting up ambushes, examining terrain in terms of its potential suitability. These veterans use weapons or fantasies of their destructive capacity in the same way that some suicidal individuals find comfort in keeping poison or a gun available.

Although Frank's fear was evident in his combat nightmares, it was important in his case, and in the cases of most of the veterans with a paranoid adaptation to posttraumatic stress, to elicit conscious emotional recollection of the terror they experienced in combat. The dramatic quality of their enraged behavior can lead these veterans, and those who treat them, to overlook the fear that underlies their stress disorder. If they are to recover, the fear must be explored and diminished in relation to both combat and civilian life.

Paranoid adaptations are characterized by a refusal to accept blame or responsibility; therefore it is not surprising that virtually all veterans with this adaptation deny guilt over their combat experiences. Those who killed noncombatant Vietnamese when it may not have been necessary often tell themselves that all Vietnamese were enemies, or that they were only following orders. In the climate of war many were not troubled and some believe they are still not troubled. But invariably they are, and acknowledging their guilt is the first step in not continuing to be imprisoned by it.

A number of common experiences can be identified in the precombat lives of veterans with a paranoid adaptation: a history of frustration or deprivation in relation to their parents, an absence of direct expressions of anger toward their parents, a history of physical aggression, and a proclivity to feel pleasurable excitement from risk-taking behavior.

If a violent response to their feelings of vulnerability was latent in these veterans, combat sanctioned the response. Although violence for paranoid veterans may relieve their sense of frightened vulnerability, providing a transient sense of excitement and potency, the consequences of such behavior are most often shame, guilt, depression, fear of retaliation, and more anxiety.

7

Mourning Never Ends: The Depressive Adaptation

AMONG persons who have experienced traumatic events, depression has been recognized as a frequent emotional response. Among concentration camp survivors, a pervasive depressive mood with morose behavior, general apathy, and a tendency to withdraw has been identified as the predominant psychiatric condition.[1]

Throughout the 1970s, a number of observers have pointed to similar responses among Vietnam veterans. Shortly after their return from Vietnam, more than a third of a large sample of Vietnam veterans studied by

Edgar Nace and his colleagues were identified as clinically depressed.[2] This depression was reflected in a general sense of incapacity which was manifested in high rates of unemployment, marital difficulties, and drug and alcohol abuse. Additional studies suggested that depression was particularly characteristic of veterans who had been in combat assignments,[3] and that it was related to feelings of guilt, either over survival, or over actions committed in Vietnam.[4]

In our experience, depression among Vietnam veterans is not always secondary to incapacity or guilt. Rather, for approximately 10 percent of those whom we have evaluated, depression—unaccompanied by substance abuse, antisocial or even suicidal behavior—is an integral part of the stress disorder. For this group, as for so many of the concentration camp survivors, the meanings of their traumatic experience are linked with the loss of close comrades, over whose death they remain in mourning.

In the early 1970s, Chaim Shatan wrote of the impacted grief that he saw among Vietnam veterans some five years after combat.[5] Today we are seeing veterans who have been mourning lost comrades for a much longer period and who often appear to have sacrificed their own vitality in the process.[6]

When posttraumatic stress is integrated around depression, and the depressive adaptation is not complicated by other obvious difficulties, it may seem to the outsider to be the mildest form of the disorder. The veteran may work, have a family, and not get into fights or trouble with the law. In knowing such veterans, however, it becomes apparent that their anguish is all the greater from keeping it to themselves. In contrast to those with a paranoid adaptation who remain tied to the war in how they treat civilian life as an extension of combat, these depressed veterans are tied to the war through their bond with those who died in combat.

Sixteen years had passed since Michael Drake had lost his two closest friends in an action in which he himself had been seriously wounded, but his mourning for them had not diminished. Michael was one of eight survivors of a platoon of thirty that had been ambushed and overrun by a large contingent of Vietcong.

Michael was a thin, blond veteran of thirty-five with a relaxed and friendly manner. He owned a successful business in a small town, was happily married, and had two children. He had been referred by another veteran for our control group as someone who was not troubled by his Vietnam experience; Michael indicated at the outset, however, that he was troubled.

From a home in which his parents fought constantly about his mother's drinking, Michael and his two siblings tried to support each other in growing up. When he was drafted in 1966, Michael formed close ties to the young men he joined in a new infantry division. He compared basic training to being in college; he and his friends would go into town together on weekends, and sometimes wrote to each others' families. Because their group was so large they went overseas together by boat, and the trip further cemented their friendships.

In Vietnam, Michael was in frequent firefights, saw men in his unit killed, and was himself directly involved in killing enemy soldiers. He said that all other firefights were "crowded out" of his mind, however, by what had happened to him in one action during the spring of 1967.

Michael's platoon had been pursuing enemy soldiers when they were ambushed. The first shots fired killed one of Michael's closest friends. The unit tried to retreat, but the North Vietnamese had cut off the escape route. Along with another soldier, Michael headed for a riverbed and dove into it, but the enemy had that covered too, and the man with him was hit in the leg. Shooting his gun in the direction of the enemy, Michael dragged his comrade out of the riverbed and told him to follow him. By this time, however, the wounded man was incoherent and took off in the opposite direction. Michael never saw him again.

Just as he approached two others from his unit, Michael was shot in the back. The other two men were also hit, but while they were pinned down they kept reassuring him that he would be all right. After a while, when he stopped hearing anything from them, he dragged himself toward them, to discover they both were dead. While he was lying beside them, artillery shells hit a nearby tree that fell down, covering him with branches.

A short time later the NVA approached. When they came to where Michael and the two others were lying, they fired two bullets at each of them. Michael heard the shots fired at him go right over his head into the ground. He also heard the NVA searching and stripping the bodies of other men, but either the branches of the tree or the pool of blood surrounding him seemed to have discouraged them from scrutinizing him more carefully.

He lay there for hours, pretending to be dead, though occasionally having to spit up blood. The NVA were only a few yards from him setting up another ambush, anticipating that other Americans would be coming to the aid of the ambushed platoon. During the night, the NVA finally pulled out, and in the morning other Americans came to rescue Michael. Recuperating later in Japan, he learned that another close friend had been killed trying to come to his aid.

When Michael returned home, he rarely spoke about his wartime experiences and because of that, felt cut off from them. He made friends, was close to his family, formed constructive business relationships, and dealt well with customers. His attitudes toward his relationships with women had a "it-don't-mean-nothing" quality, similar to what he and his friends often said of their experiences in Vietnam. After ten years, when he started dating the woman he later married, he felt somewhat differently. He described her as interested in and willing to talk about his experiences in Vietnam. Yet he told her he would "drive her crazy" if he talked about Vietnam whenever he was thinking about it, which was for extended periods every day.

Each year on the day before the anniversary of his being wounded, Michael went over everything that happened to him in Vietnam. Year-round, he would dream of Vietnam at least once a month and have a nightmare about his experiences every few months. In his nightmare there would be hordes of Vietcong coming toward him and the other American soldiers. Sometimes he would be firing his gun and mowing enemy soldiers down, but more and more of the enemy would continue to come. He would be apprehensive that he could be shot at any moment, although in the dream he never was.

Michael always made sure when he fell asleep that his arm was not under his body in the position it had been during the episode when he had pretended to be dead. A couple of nights a week he would wake up, unable to go back to sleep for several hours.

Most striking about his posttraumatic stress symptoms was Michael's preoccupation with the two friends who had died in the ambush. He dreamed of them frequently, but when he did they were never dead. The three of them would be together, doing the things they used to do, kidding around, and having fun. If he did not dream about his friends for several weeks, he would miss them and want to dream about them again.

When Michael first came home he would often cry while watching the television news about Vietnam and sometimes remarked to his father that he should be dead along with his friends. Once, soon after coming home, he drove five hundred miles in a snowstorm to see one of his friend's parents. He continued to keep in touch with both of their parents, calling them on Christmas and other special days.

During the day, Michael's thoughts frequently drifted to his dead friends. Whenever he saw groups of teenagers, he remembered how young his friends had been and the thought exacerbated the pain of their loss. The feeling that everything was going so well now in his own life dramatized how much they had missed out on.

Wishing to reunite the men who had survived the ambush with him, Michael located several from his former unit and met them in Washington on the dedication day for the Vietnam Memorial. Unlike veterans who were able to use this occasion to mourn the death of friends, accept those deaths, and go on with their lives, Michael was determined to continue his mourning.

But despite his preoccupation with Vietnam and the loss of friends, Michael was able to enjoy his family and gain satisfaction from his work. He did not see himself as needing or wanting help. For other depressed veterans with posttraumatic stress, however, pleasure and enjoyment seem to be impossible. Many of them turn out to be mourning not only lost comrades, but also what was best and most alive about themselves. In a psychodynamic sense they are mourning their own deaths.

The price paid by Jack Edwards in his lonely struggle with the aftermath of his Vietnam experiences was so high that it barred him from enjoyable participation in the rest of his life. Jack was a thirty-six-year-old Vietnam veteran, married, with three children. He was tall, slender, soft-spoken, articulate, and markedly depressed. His lack of interest or enthusiasm for anything dominated his initial interviews. He had not worked for several years and was receiving a disability pension from the Veterans Administration for wounds sustained in combat.

Jack was the oldest child of a religious Catholic family who lived in a working-class suburb. He was a responsible youngster who did well in school, worked part-time, and helped look out for his sisters and brother. He respected his parents and accepted their authority. His mother died when he was fourteen and he, his father, and siblings went through a difficult emotional period for the next year. His father drank heavily, became withdrawn, and was barely able to function at his job in a local factory. Jack withdrew from social activities and sports for the year following his mother's death, but resumed them when he was fifteen. He was a good high-school athlete and a successful cross-country runner. After finishing high school in 1965, he enlisted because he believed in President Kennedy's Vietnam position. Jack knew he would be a capable soldier and felt it was his duty to serve.

Wanting to be in the paratroops, Jack joined an airborne division. He was sent to Vietnam in the spring of 1966 and remained there until May 1967. Most of the men in his outfit had trained together in the States, and since he felt close to his group, it was painful for him when someone was wounded or killed. Jack's outfit was mainly involved in combat assaults in an area outside Saigon, which Jack described by saying, "When you went in you always made some contact with the enemy, and since they saw you coming they always struck the first blow."

On most of his unit's operations they landed by helicopter, which always frightened Jack because of the uncertainty of what would be encountered on the ground. Although he had been trained as a medic and was assigned that duty, he also played an active role in the fighting, killing many enemy soldiers and being wounded several times.

One of Jack's most painful experiences involved retrieving a wounded friend from a rice paddy. As he was pulling him out, Jack realized his friend was dead, but only when he heard shouts from other soldiers did he realize that his friend had been severed in half, and he was pulling out only the top of his body.

Most of the killing Jack did was from a distance and seemed impersonal. The first time, however, the experience was different. As Jack's unit went into a village, an old man with a beard came running toward them with a grenade, and Jack shot and killed him. A middle-aged woman whom Jack assumed must have been a member of the old man's family came out, quietly crying. He felt that the old man should have died in bed with his family around him, rather than been killed by an eighteen-year-old kid. The men in his unit congratulated him because he had prevented them from being killed, but Jack felt more sadness than pride.

He retained his sense of compassion throughout the war. Whenever he passed a wounded North Vietnamese soldier and he had time, he would stop and take care of him. In one instance, by making a joke of the incident, Jack prevented his unit from killing a North Vietnamese soldier who had thrown a grenade at them. He said they had been taught in training to attack and kill, but that he had been more deeply impressed by what he was told as a medic: "It is easy to kill, but hard to save a life."

Toward the end of his tour, a bullet fractured his right ankle and he was evacuated to the States for treatment. Back home, Jack felt guilty about leaving his friends overseas, felt cut off from his family, and was uncomfortable with civilians who were opposed to the war. Recuperating from his injury, he learned that the man who had been his closest friend in his unit had been killed. Jack felt his friend had become more vulnerable as their unit was now comprised primarily of inexperienced replacements. Despite the fact that he had just married a woman he had known before going to Vietnam, Jack reenlisted. He wore an elastic bandage over his wounded ankle, told the Army doctors it was due to a recent sprain, was accepted and sent back to Vietnam.

During his second tour Jack was initially stationed in Saigon, but requested and was given a combat assignment. With four other Americans, he was assigned to a unit with 125 South Vietnamese to perform search-and-destroy patrols in the Delta area. Normally the four Americans rotated going out on the patrols, but Jack would often take others' turns as well as his own because he felt he was more experienced and he also wanted to be in the thick of the fighting. He was eager to win medals and would lead the attack at every opportunity. He received several Army combat medals and a Bronze Star.

After completing his second tour in late 1969, Jack went home on leave, planning to return to Vietnam. By that time, however, his wife had had a child and she persuaded him not to go back. He reenlisted instead for an overseas assignment in which his family could be with him; two more children were born overseas. Jack remained in that assignment until 1975, when he left the service.

His first child was born with a deformed arm that the doctors said was extremely rare and unlikely to happen a second time. His second child had no defects, but the third was born missing a foot. Jack was constantly pained and saddened by the taunts his two children had to endure from their classmates and by the limitations their handicaps placed on them. He wondered whether Agent Orange, which he was exposed to in Vietnam, had something to do with their birth defects.

When Jack returned to the United States, he became a car salesman. At first he was enthusiastic and successful enough to be promoted to a managerial position. A combination of physical symptoms, aggravated by the strain on his injured ankle, and gradually increasing symptoms of posttraumatic stress disorder culminated in his leaving his job in 1978. Jack was ashamed of the fact that his wife worked while he did not, and for many months in therapy did not reveal that her income was their principal means of support. Because he did not want his children to know he was not working, he pretended he was in business. He also told them that he had not participated in combat, and avoided any discussion about their congenital deformities possibly being related to his Vietnam experience.

A great deal of Jack's time was spent thinking about Vietnam and his friends who died there. Not a week went by without a nightmare dealing with the war. In one of his most frequent recurrent nightmares he is turning over the bodies of soldiers who had been killed, something he frequently did in Vietnam. In the dream one of the bodies is his own and the body is rotting. Another nightmare replayed the incident in which he had pulled his wounded friend's severed body out of the rice paddy.

Jack was emotionally removed from people when he came back from combat, and his sense of estrangement grew worse as the years passed. He was often bored with people and felt increasingly removed from his wife and children. He was seldom able to feel pleasure or excitement; he had insomnia, an intense startle reaction, and experienced exacerbation of his symptoms in response to anything that recalled his war experiences. Since being in Vietnam he could not eat meat without thinking of how it had been killed.

He also felt considerable guilt over the killing he had done, noting that "every life is irreplaceable." He felt that most of those he had killed were not Communists, rather, "They were kids who did not know what Communism was any more than I did at that time." The enemy soldiers he fought in Vietnam were usually young teenagers, but since Jack was only eighteen when he first arrived in Vietnam, he had not thought too much about it. When his oldest child approached adolescence, however, he became conscious that he had killed people of that age. Whenever he read of parents who had lost their children he identified with the pain and loss of the parents of his friends who had been killed in the war. These incidents also exacerbated his personal pain over what had happened to his children.

The initial evaluation of Jack's posttraumatic stress disorder focused on his depressed mood, his isolation, and his lack of pleasure. His recurrent dream image of finding his own dead body seemed to represent his sense that part of him had died in Vietnam. He appeared to be in mourning for himself, as well as for his friends who were killed. His deadness and withdrawal seemed to be a way of minimizing the difference between their death and his life.

As his treatment continued on a once-a-week basis over several months, the connection between Jack's sense of isolation and his combat experiences became clearer. Early in therapy, he had a dream that he said he had had many times since his return from Vietnam: "I was trying to get out of a leper colony and the lepers were keeping me prisoner."

Discussing the dream, Jack said that there had been a leper colony not far from where he was stationed on his last overseas assignment, and that he had once been unofficially involved in arranging to ship some discarded Army supplies to the colony. He also spoke of a movie he had once seen about a leper colony in which the lepers were regarded as "unclean and contaminated." He had several times previously used the word "unclean" to describe his own postcombat perception of himself. He related this feeling to killing, saying, "Killing both makes you unclean and sets you apart from others."

Jack said that half his problem was how others reacted around him, knowing that he had been a combat soldier and had probably killed people. He once heard someone at a party mention in a whisper that he had been in Vietnam, "with all the implications that seemed to carry." He said he could understand how others felt, since he was uncomfortable around people who lied or stole—and killing was a lot worse.

In subsequent therapy sessions Jack elaborated on this theme. He compared killing to the loss of innocence that sets apart the first boy in the class to have sex. He explained the lengths he went to in order to avoid letting his children know about his combat experiences because he thought the knowledge would take away their innocence. He viewed his Vietnam experience not only as a disease, but as a contagious one from which his family had to be protected, though it was clear there was no way they could be.

Although depression was Jack's paramount symptom, a readiness to fight was also part of the picture. He related several incidents in which he had been attacked or provoked and had either fought or intimidated the other person involved. When a young man coercively asked him for money in the park, Jack pretended to reach in his pocket, but instead punched the man. When a man on a train insulted and tried to intimidate him, he knocked the man to the floor. When people behind him and his wife in the movies talked incessantly, he reprimanded them, with an implication in his tone and manner that if they said anything more they would be sorry. Jack did not carry a weapon, however, and seemed not to fear that he would seriously hurt anyone.

These incidents indicated a heightened sensitivity to potential attack, with some of the same readiness for quick physical response that we have seen among more paranoid veterans. In this context, Jack related that while in Vietnam, if he heard something suspicious in the bushes he would just throw a grenade. The infrequent reflections of combat behavior in his civilian life were interruptions of—and perhaps provided some relief from—the more pervasive depressed mood Jack showed in his postwar adaptation.

Jack's behavior in combat and civilian fights contrasted with frequent dreams of feeling helpless and overpowered in a combat-like situation. For example, he dreamed that he was firing his gun at a North Vietnamese soldier coming toward him and he kept missing. Finally the soldier was on top of him and Jack hit him hard with the butt of his rifle, but the soldier just smiled as though he had not felt the blow. Then they were fighting and Jack had to use both hands to prevent the soldier from using what looked like a sabre against him. Nothing he did to the soldier seemed to have any effect.

Jack remarked that in none of his dreams was there ever any "outcome," adding that even when he had a sexual dream, "it ends before anything happens." He related his dreams to his sense of impotence and ineffectiveness in civilian life and his lack of enthusiasm and pleasure.

Although devoted to his children and protective of them, Jack felt their handicaps produced a sadness that prevented him from enjoying them. The sadness extended as well to his middle child, who was not handicapped. Yet all his children were cheerful in disposition, and had active social lives. Jack felt they took after his wife, whom he considered a cheerful and optimistic person. At the same time, he felt that after fifteen years their relationship had evolved into a close friendship in which there was no excitement, sexually or otherwise.

In civilian life Jack seemed to be checking the assertiveness and potency he had demonstrated in combat. His analogies of the separateness and loss of innocence conferred on a young man by sex and by killing implied that his sense of the "uncleanness" of killing derived in part from the excitement connected to it. Jack's loss of interest in work, his lack of sexual pleasure, and his general loss of enthusiasm was related to the combat-derived linkage between effectiveness, excitement, potency, and destructiveness.

The relation between his lack of pleasure and his need for vigilance became more evident when Jack bought his wife a radio that did not work properly. Jack anticipated an argument with the salesman about returning it. That night he dreamed he was in the store listening to the radio playing beautiful music. Then the radio stopped working and he firmly persuaded the salesman he had to exchange it. Then he became aware of a fire in the building; he had been so absorbed listening to the music that he had not noticed the smoke. Suddenly he was on a ledge of the building, five stories up, with his wife and children. He took the whole family in one swoop and jumped toward a safety net below.

Jack connected letting down his guard—his involvement with the music—to a conflict between pleasure and awareness of danger. In Vietnam he had always expected the worst to happen and that was how he had protected himself. The fire, however, related to his anger and appeared to confirm that he equated his effectiveness with the salesman with destructiveness, much as he had in the war.

Jack went on to speak of the need he felt to be perpetually vigilant in protecting his family, particularly his children. He constantly feared they would be injured, run over, or attacked. He was beginning to recognize that his own mood was perhaps the greatest source of danger to his family.

At this time Jack started to miss sessions, giving credence to the therapist's suspicion that the dream of the radio that no longer played well was also an indication that he no longer liked the sound of what was transpiring in his sessions. When asked about this, he confirmed that challenging his isolation had been making him feel increasingly uncomfortable. He felt he was being encouraged to share something of his experiences in Vietnam with his family, but he was reluctant to do so and, rather than hurt the therapist's feelings by admitting that, he had decided to keep quiet about it. He had begun to be overprotective of the therapist, as he was with his family, and in the process, was keeping an emotional distance much as he did with them.

Shortly thereafter, Jack related a dream in which he was a prisoner in some kind of cell without guards; he had the feeling it was in a foreign country. In the dream, he was aware that if he were going to be there for four or five years he would miss watching his children grow up. Jack associated being cut off as a prisoner in the dream to the way in which he was a private person, not showing other people how he felt.

He was uncomfortable, for example, when his children wanted to see pictures of his family. He did not want them to see his father in uniform, fearing they would get the idea that military service was a family tradition and that they too should enlist at a young age. He also did not want to have to share the pain of his mother's early death with them.

It was pointed out to him in therapy that while he had described himself as close to his sisters and brother growing up, he had not spoken about his postwar relationships with them. He revealed he was disappointed and pained by what had happened to each of them. One of his sisters had cut herself off from the family; the youngest sister had been heavily involved with drugs; and his brother had gone to live in a religious commune. Talking about his siblings comprised another part of Jack's discomfort about showing his children the family album.

Jack viewed not only Vietnam, but his entire past as something that could harm his children and strove to protect them from it. In essence, he was protecting himself. In the process he cut off an empathic connection between himself and his children and the excitement of their growing up. His dream about the prison cell was an acknowledgment that he perceived himself in his own home as a foreigner, a stranger, and a prisoner of the Vietnam War.

In later sessions Jack revealed that the relationship between his parents had also been more problematical than he had initially indicated. His mother's parents had never accepted his father and constantly fought with him. His mother and father would fight over her defense of her family, and his father would at times hit his mother. As an adolescent, Jack had frequently interceded to protect his mother; he was sympathetic to his father's feelings, however, because his mother's family also mistreated him and the other children.

In response to questioning in therapy about his need to keep his children distant from his life, he related a combat experience that seemed to confirm for him his idea that having killed made him a killer. He had been part of a squad of seven men that came upon about fifty Vietcong who were sitting, eating, and relaxing. The squad opened up with grenades and automatic rifle fire and killed about thirty of the group. Jack described the Vietcong as young kids who behaved courageously. They did not scream, but looked at each other in surprise and went for their rifles. After the attack Jack's colonel congratulated the squad, but all of them felt lousy about it. Jack described the incident as "like shooting fish in a bowl." He remembered lying down exhausted that night, but being unable to sleep.

Following the retelling of this episode, Jack had a dream that a large American city was under siege. He and some other men were in an underground tunnel with M-16s. They were being attacked on all sides, but the enemy did not have automatic weapons so Jack and the others were killing whoever shot at them. He asked someone next to him what the sense of all this was and how long it was going to go on. The person replied that it was better to be shooting than to be shot.

The dream stimulated Jack's memory for more of the details of the experience of killing the Vietcong. He recalled the fear of being seen standing up to throw grenades at the group. As he threw them he could see the seams of the grenades; each one had seemed to take forever to land. He recalled the difficulty he had seeing with all the dirt flying. With his first magazine clip he missed hitting anyone, but got off two more rounds. He said again in this session that it was like shooting fish in a bowl.

The dream reflected that for Jack the war was still being waged. Although the superior firepower and the multiple killings paralleled his combat experience, Jack changed the one aspect that bothered him most: surprising and killing men who had not attacked him.

Jack insisted that the war had changed him into someone who would without compunction react to danger with excessive violence. Despite the fact that his civilian fights were not characterized by uncontrolled violence and he did not carry or even own a weapon, his perception of himself was as a killer. His dream dialogue, about it being better to shoot than be shot, was a reflection that he was beginning to question this perception. Moreover, in this dream, unlike his earlier combat dreams, he was effective. At this point in his treatment, Jack became less depressed and began to talk of the possibility of returning to work.

The work question took Jack further into his concern about what was owed to the dead and to their survivors. While he had been in basic training a friend had been killed in an automobile accident. Jack had been assigned to escort the man's body back home for burial. Since the war, the friend's father, a wealthy hotel owner, had on many occasions offered to give Jack a job.

Jack felt to take such an offer would be to profit from his friend's death. He also felt it incumbent upon him to hide the congenital deformities of his children from his friend's father, as the man had suffered enough pain and he would not inflict more. He could not accept the idea that helping his son's friend might have made the man feel better. Jack's reaction to his friend's death had been to refuse to learn to drive—now even as a passenger he was extremely tense and uncomfortable in a car.

Not driving and not working were part of the sacrifice Jack's grief required. A tragic event brought his need to preserve his grief into sharper focus. A veteran friend from Jack's company in Vietnam came to town for a few weeks and spent a lot of time with Jack. The friend had been working in Puerto Rico and had fallen in love with a young woman who lived there. There had been problems in their relationship and when he went back to Puerto Rico, he killed the girl, shooting her many times, and then committed suicide.

For the first time since the sessions had started, Jack called the therapist at home to tell what had happened. During his friend's visit Jack had found him to be something of an emotional drain and was upset about that. He was also angry at a mutual friend who was not taking off from work to go to the funeral, which was to take place in a nearby city. In a session shortly thereafter he continued this theme, using events like the funeral to justify not working and claiming that when caught up in work people lose their values and perspective.

When Jack went to the funeral, he had not told his wife of his friend's suicide and refused to tell her where he was going. He thought he wished to spare her pain, but in the process of not sharing his grief he was nursing it, making sure it did not dissipate. About this time he dreamed he was a prisoner with his friend and they had to crawl through underground tunnels in which they could hardly move or breathe. The tunnels suggested to Jack both his real experience in tunnels in Vietnam and a coffin.

Jack's reaction to his friend's suicide was a reenforcement of reactions he had had to the deaths of friends in Vietnam and heightened his determination to dedicate his life to mourning. Anything that interfered, whether work or pleasure, must be resisted. The dream reaffirmed his identification with those who had died.

Yet more than any discussions of his combat experiences, this event made it possible for Jack to eventually see how he had sacrificed his life to grief. His depression, which had returned in full force after his friend's suicide, gradually diminished. After five years of not working he went into business with a friend and was remarkably successful. His combat nightmares became less frequent and he was no longer powerless in them. When he was under attack in dreams by North Vietnamese regulars, he and his squad repulsed them. After a year of treatment, he had made satisfying progress, but the outcome was yet to be determined.

More than a shattered capacity for effective action was suggested by Jack's case. He and many veterans like him were effective for years after their traumatic combat experiences. Jack's effectiveness and success appeared to give him the feeling that he was putting Vietnam behind him, something he was not prepared to do.

Guilt over killing and over enjoying life when death had deprived his friends of that possibility certainly contributed to Jack's reduced functioning. But it was his bond to his dead comrades and the feeling his life had ended when theirs did that gave Jack's depression its distinctive character.

Some factors in Jack's precombat history are particularly relevant to his situation. His mother had instilled in him a strong sense of values regarding church and country. His highly developed ethical sense was the source of his strength, his compassion, and also much of his pain. His values also represented an ongoing tie to his mother.

It is difficult to assess how much the grief and depression Jack experienced over his mother's death before the war made him vulnerable to a depressive adaptation to posttraumatic stress. It seems to take on more significance from our observation that pain over a mother's death, or over abandonment or rejection by her, is a dominant feature in veterans whose posttraumatic stress disorders center on grief and depression. At the same time, since we have seen veterans who experienced severe maternal rejection in childhood, but dealt with combat without such a depressive reaction, indeed in some cases without developing a posttraumatic stress disorder, the link between the two factors is not of a strictly causal nature. What is clear, however, is that veterans whose posttraumatic stress is complicated by a depressive adaptation are aware of and talk of their earlier pain, loss, or rejection in a way that veterans who did not develop such a reaction do not.

Some veterans, like Michael Drake, who were hurt by early family experiences related to other men in their units as if they were brothers in a new and better family. Both support and vulnerability were inherent in such relationships, but prolonged mourning reactions seem to take place in men who were already vulnerable. Veterans with depressive adaptations, like Jack and Michael, have prewar histories reflecting sadness rather than, as is more often seen in those with postcombat paranoid adaptations, histories in which pain is dealt with by fighting or drinking.

With many veterans the difference between a depressed and paranoid adaptation is clearcut, but in others the picture is more mixed. We have found it helpful to make the distinction between the two adaptations and to understand the different psychology that goes into both. The paranoid response involves a perception of combat as never ending, a perception that requires vigilance and counterattack to survive; the depressive response reflects a perception of irretrievable loss, a triumph of death over life.

If war had broken out when he was first in therapy, Jack, like many other veterans suffering from the aftermath of combat, would have been eager to participate. Given the pain and suffering of the Vietnam experience, a wish to return to it may seem puzzling. The desire to go back to an experience that was so traumatic can be perceived, however, as combat nightmares have been, as an attempt to master an experience which remains overwhelming.

The desire to return includes more than a need to complete unfinished business. For veterans with a paranoid adaptation, like Frank O'Donnell in chapter 6, the return signifies an attempt to recapture the excitement of combat, an excitement that served to help them master their fears in a manner that finds its parallel in the excitement of their fights in civilian life. For veterans like Bill Clark (chapter 5), whose guilt and need for punishment are primary outgrowths of combat experience, their unfinished business in Vietnam seems to be the death they feel they deserve. For veterans in mourning, like Jack Edwards, the fantasy of a return to Vietnam includes a seemingly contradictory way of holding on to their grief and the hope that they might be able to recapture their own lost life. The dream in which Jack is turning over bodies and discovers one is his own, expressed his wish to recover the dead part of himself, the vitality that perished in Vietnam. The fact that even depressed veterans like Jack Edwards come to life discussing their combat experience, expresses the intensity of combat's therapeutic meanings and the power memories of it have as a therapeutic stimulus.

Complications

8

Crime

ADISTURBING legacy of the Vietnam War is the large number of men who have returned from combat to civilian lives marked by crime. Since the late 1970s law enforcement officials have been aware that a disproportionate number of Vietnam veterans have been involved in criminal activity.[1] In 1978, a Presidential Review Memorandum estimated that 29,000 Vietnam-era veterans were incarcerated in federal or state prisons, and approximately 37,500 were on parole, 250,000 under probation supervision, and 87,000 awaiting trial.[2]

The role of combat in encouraging violent or antisocial postwar behavior has not been well understood. Some observers have been struck by how guerrilla warfare in Vietnam led many soldiers to overcome and neutralize earlier resistance to violence.[3] The continuation of this stance in civilian life was suggested by a large-scale study of the early 1970s which reported that Vietnam veterans who had experienced heavy combat approved the use of violence to a significantly greater extent than did non-veterans.[4]

Yet the overwhelming majority of soldiers who fought in Vietnam returned, as in previous wars, to peaceful law-abiding lives. This has led some to contend that the incidence of criminal behavior among veterans of the Vietnam War has been exaggerated and sensationalized by media reports, and especially through fictionalized television dramas.[5]

While sensational media reports certainly do not present an accurate picture of all Vietnam veterans, they do, however, underscore a phenomenon that appears to be widespread among the high proportion of veterans who are suffering from posttraumatic stress disorders. Our studies, which show that almost 40 percent of this group have engaged in criminal behavior during the postcombat period, indicate a relationship between combat and postwar crime.

A striking feature of veterans with posttraumatic stress who have engaged in postwar crime is the very large proportion—almost three-quarters—who evidenced excessively violent or criminal behavior while in Vietnam. More than half of these men had no precombat history of violent or antisocial behavior. Yet during combat, excessive brutality against the enemy in the form of mutilations or torture, as well as personal violence against civilians or persons not known to be the enemy, was widespread among these particular veterans.

More than one-third of the veterans we have seen who have committed crimes in the postwar period participated in "fraggings," or other attempts on the lives of other Americans in Vietnam. Analyses of violence that American servicemen inflicted on other Americans have shown that this behavior—which appears to have been far more frequent in Vietnam than in any previous war—became particularly pronounced during the later years of the war (1970–1972). This increase has been attributed primarily to the generally low morale among American troops, increased use of drugs, reduced level of combat encounters, and the prevailing sense of purposelessness that characterized the last years of the American involvement in Vietnam.[6]

The nature of veterans' most frequently committed crimes reflects a strong proclivity towards violence. Leading the list of crimes of men we have studied are assault and battery, assault with a deadly weapon, armed robbery, possession of a concealed weapon, and homicide, both attempted and completed. About 5 percent of the veterans we have seen have been involved in organized criminal activities, including contract killings, extortion, and racketeering. Less frequently reported are crimes against property: burglary, petty larceny, grand theft, and shoplifting. In only one veteran have we seen so-called "white collar crime," in this case embezzlement of a large sum of money from the bank in which the man was employed.

Although for the larger proportion of veterans we have studied with postcombat criminal activity, the roots of criminal behavior first appeared in Vietnam, the rest of this group—about one-third—had shown antisocial behavior before combat. Some of the most sadistic combat histories in Vietnam involved the latter group of men, who were sometimes given assignments in which their antisocial tendencies presumably were an asset.

Many have wondered whether stress disorders would be found among such antisocial individuals.[7] Were men who seemed devoid of conscience, or capable of killing without evident remorse, and who often appeared indifferent even about preserving their own lives, protected by their own hardness and insensitivity? Our work indicates they were not. Some may have killed noncombatants with indifference—indeed killing conferred on them a sense of power and invulnerability—but when that illusion was shattered, as often resulted from the circumstances of war, they were no more immune to terror than anyone else. If anything, previous unnecessary violence made them feel, if not guilt, a fear of retaliation that exacerbated their stress disorders.

Finally, we have found that one-quarter of the veterans involved in criminal behavior since their return from Vietnam had no record of excessively violent or criminal precombat or combat behavior and became involved in such behavior only after developing a posttraumatic stress disorder. Our work supports the observation of others that antisocial, criminal acts that are not the result of preexisting criminality are one of the major deviations through which posttraumatic stress may be manifested.[8]

The varied dimensions of crime among veterans with posttraumatic stress are reflected in the two cases presented in this chapter. One veteran had a long history of antisocial and criminal episodes prior to the war; the other showed no signs of violence or crime until after he returned home.

Before the Army, Ray Thompson had been involved in burglary, loan sharking, and auto theft; had spent two years in prison; and had to get a special waiver for acceptance in the service. In Vietnam he was recruited to be in charge of a South Vietnamese Army team that regularly tortured, mutilated, and killed villagers in attempts to obtain information about Vietcong operations. Although Ray initially permitted others to do the killing, he was increasingly excited watching it, and soon became actively involved. He killed a mayor in front of the people of his village; mutilated a woman and her baby in an effort to get her to reveal where her husband, a suspected Vietcong leader, was hiding (she eventually revealed the hiding place, but too late to save her life or that of her child); and ordered and participated in the massacre of all the inhabitants of a village suspected of harboring Vietcong.

Ray was excited by combat as well. The one time he killed an enemy soldier in hand-to-hand combat he was sexually excited by the experience. He prided himself on the fact that during firefights he would always order the South Vietnamese troops in his command to move forward and attack. Years later he expressed remorse that he had needlessly lost men in this way.

During Ray's sixth month in Vietnam, the Vietcong wired a young boy with a grenade, and the child blew himself up a few feet from where Ray and some other soldiers were standing. Ray was seriously wounded and two South Vietnamese soldiers were killed. For the remainder of his tour Ray lived in terror. At the same time, he came to see his injuries as a punishment for past behavior, was preoccupied with guilt, and began drinking heavily. Although before the grenade injury, his violence and his control over who lived and who died had made him feel powerful and omnipotent, he now felt vulnerable and unable to protect himself.

By the time Ray returned home, he had developed a posttraumatic stress disorder centering on both guilt and fear. Once, when a young Oriental boy walking toward him on the street caused him to relive the experience in which the grenade had gone off, he dove under a car to protect himself and was slightly injured. His stress symptomatology was insufficient, however, to prevent involvement in criminal activity that included contract killings for an organized crime syndicate, which recruited him partly because of his wartime experience. Ray, who had become an addict, was evaluated but not treated by us. We later learned that he had died of an overdose of heroin.

The horror with which one is likely to respond to the combat and postcombat behavior of men like Ray Thompson should not distract us from the fact that if combat can undermine the values of previously ethical young men, its corrosive effects on those who are already antisocial, or marginally so, is apt to be devastating. The tragic impact of war is even more boldly underscored, however, when one witnesses, as in the case of the next veteran, the destructive transformation of young men who seemed destined for better things prior to Vietnam.

Warren Saunders was happily married with two children and a good job when he volunteered for service in 1965. He had heard that the war in Vietnam was a small "police action" with little risk, and planned to use military benefits to complete his college education. When we first met him in 1982, after his distinguished tour in Vietnam as a helicopter pilot for which he received a Presidential Citation, a Valorous Unit Award, fourteen air medals including one with a "V" for heroism, and a Purple Heart, he was a prison inmate serving a two-to-six-year term for bank robbery.

Warren was a tall, well-built man with reddish hair who seemed depressed. At times he quietly cried when talking of men he had known who were killed in combat. At other times, he became anxious and apprehensive, particularly when discussing situations in which he had almost been killed. He was intelligent, articulate, and related his combat events and the various episodes of his postcombat life in great detail.

Warren grew up in a rural area in the northern part of New York State, and was the fourth of six children. His father worked for the township, plowing snow in the winter and fixing roads in the summer. His early memories of his family were warm and happy, centered primarily on being with his mother in his pre-school years, keeping her company when she did her chores, and accompanying her when she went out. His memories after six included his father and older brother, with whom he went hunting and fishing.

Warren saw his father as an easy-going person; his mother was more of a planner and organizer. From the time he was very young Warren helped milk the cows and take care of the animals at the house. In school he ranked just below the top in his classes, and said that had he done better he would not have been liked by his friends.

After graduating from high school, Warren married a young woman he had known for several years, and in a short time they had two children. He described his relationship with his wife and the children as warm and loving. At the suggestion of his older brother, who was in the banking business, he went to college to study accounting, but an auto accident caused him to be out of school for several months and he never went back. For a while Warren went into business selling plumbing parts with a friend, then later took an administrative position in an office before enlisting.

Warren was trained as an Army pilot, became a warrant officer, and from August 1966 to August 1967 served as a helicopter pilot in Vietnam. He flew in the southern part of the country, sometimes flying into Cambodia. In this assignment he frequently flew Special Forces units on missions in which he dropped them into particular areas and picked them up several days later. During the last three months of his tour he was made a section commander and flew combat assaults in the northern plateau region. He saw extensive combat and was wounded once.

Warren's first traumatic combat experience was in September 1966. He and another pilot returned to pick up five Special Forces soldiers after receiving a message that their mission had been compromised. There was no clear area to land, however, and they were under heavy fire. Warren's helicopter went in first and dropped a ladder from each side to the men on the ground. But the helicopter was defective in pulling power and lost its upward thrust, settling onto the top of a tree with two men suspended on the ladders, a sitting target for ground fire. Warren's helicopter was blocking the only opening in the trees so the other helicopter could not get close enough to drop ladders to pick up the remaining men. The radios were open and they could hear the three men on the ground screaming that they were running out of ammunition. A fixed-wing plane was called in to spray the area with napalm which drove away the enemy, but singed the men on the ground. Meanwhile, the men who were suspended on the ladders managed to climb up to the helicopter. It was unable to take off, however, when a cable became stuck on the tree. Lacking any tools to cut the cable, one of the men shot it free with his gun, and the helicopter was finally able to lift up. The other helicopter then was able to pick up two of the men on the ground. Warren's helicopter got the last man out by throwing him a rope. Relating this incident in therapy, Warren was primarily depressed, but expressed anger about the defects in his helicopter and equipment.

Later that same month, Warren's helicopter and four others were ordered to fly a mission into Cambodia. He said he had felt then that the major in charge should not have ordered the mission since it was almost night and the weather was bad. The helicopters could not find a place to land and by the time the major gave the order to return to the base, one of them was out of fuel and crashed. Warren became tearful recounting how the pilot had radioed him to tell his wife that he loved her right before he went down.

On the way back that night Warren got caught in a thunderstorm. Ice formed on the wing and the rotor blade stalled several times as the helicopter was sucked up and down at the will of the storm. His copilot, who was a far more experienced flier than Warren, started to scream and cry. The base could not pick them up on the radio, the instruments failed, and Warren recalled pleading, "Somebody, please find us on the radar." Finally, he made contact with a ground unit and had them fire flares to guide him down.

Later that fall, Warren had the one experience of knowing for sure that he had killed someone. Landing on an assault mission, Warren and his copilot were fired on by an enemy soldier directly in front of the helicopter. Since his machine gunner could not get his gun into position, Warren reached out with his pistol and killed the man. Because he had to let go of the controls while he shot, the helicopter wavered. Warren recalled the other pilot was angry about this afterwards.

In November 1966 Warren learned that his wife had given birth to twins. Three days later he learned that one of the twins had died. He was given a thirty-day leave to go home for the baby's funeral. He cried as he related this, saying how much he liked children, how he had seen too many people die, and that it just was not fair.

In February 1967 Warren flew another traumatic combat assault mission. The soldiers he was carrying got out on the left side of the plane, but the lieutenant in charge got out on the right side and had to cut in front of the helicopter to join his men. The lieutenant's head was smashed by rotors of the helicopter, his arm was completely severed, and blood was splattered over the windshield of Warren's plane.

In the early spring of 1967, Warren was part of a company of twenty helicopters involved in a major operation in the Chu Lai area. The Vietcong evidently expected them and had mined the area; nine of the twenty planes involved in the operation were lost. Warren had been feeling ill that day and a friend who flew in his place in the first wave of the assault was killed. The situation became so grave that, despite being ill, Warren flew into the area eight times that day, bringing in reinforcements.

During that operation Warren's tail gunner was shot and killed, and a number of the men he landed were killed immediately after getting out of his helicopter. He related a particularly gruesome memory of seeing late that night, in the light of exploding shells, a soldier's severed head lying just outside the helicopter door. The head's mouth was open, as though the man had been screaming when he died and his expression had frozen on his face.

During this mission Warren carried back wounded and dead; in one day he transported 171 dead bodies. The mission was considered a success, however, since after securing the area the Americans killed 1,500 of the enemy. He spoke of the mission with considerable anger, particularly over having been shot at, and because his tail gunner and his pilot friend had been killed. The pilots felt that their security must have been compromised since they had been expected by the enemy; they blamed this on the Vietnamese civilians whom the Army Air Force command allowed to roam around the base.

On one of his trips back to the base during the mission Warren spotted distress flares and stopped to pick up a wounded American soldier and a medic. The wounded man was screaming and grabbed Warren's seat harness, crying over and over, "Mommy, Billy is dying." The man did die and afterward Warren noticed that he had a tiny opening in the back of his head where a bullet had hit him.

In a subsequent mission a short time later, Warren was wounded. He had dropped men off and was about to leave when a woman came out of a hole in the ground and fired through the helicopter at him. Warren was wearing a protective metal shield and a flak jacket, but credited the gun he wore next to his groin with having saved him. He was bruised in several places, had shrapnel in his wrist, his gun was bent out of shape, and he had black-and-blue testicles. His previous command pilot had taught him the trick of wearing the gun over his groin, saying that if he got shot there he would never go home, but would just stay until he got all the Vietcong. Warren laughed, recalling the black humor of that story.

During a landing later that spring an elderly woman shot at Warren point blank with a shotgun containing metal nails. He remembered trying to shrink behind the metal plate in the plane until soldiers on the ground shot the woman. Around this same time he witnessed the results of a brutal massacre when he had to fly American officers to a Montagnard village. Thirty-five or forty men, women, and children had been killed and had then been cut up by the Vietcong. Speaking of this incident, Warren became angry and depressed.

Although Warren said that before the war he had been totally faithful to his wife, after a few months in combat, he began to spend much of his free time with prostitutes and other Vietnamese women. During much of his tour he was involved with one particular tea-girl, whom he described as "not really a whore." A Canadian man had made her pregnant, then had left her to support herself. When the baby was born, Warren listed himself as father of the child. Although he said he had loved this woman, he had other involvements as well; his Vietnamese girlfriend called him a "butterfly" because he was not able to stay in one place. He said some men coped with the war by using drugs or alcohol; he used sex the same way.

While he was still in Vietnam, there were indications of the posttraumatic stress disorder that Warren was to develop when he got home. One morning he could not be roused for flying duty and cursed the men who tried to wake him. He woke up several hours later with no memory of the incident. This episode occured the day after Warren had learned that the wrecked helicopter and the remains of the men who had crashed when they ran out of fuel had been found. Until then it was not certain what had happened to them. The incident was not repeated and Warren continued to fly, but he suffered from severe insomnia during the last few months of his tour.

After his return to the States in the fall of 1967, Warren spent a year as an Army flight instructor in Georgia. He had come to love flying and had decided to make a career as an Army pilot. He found teaching difficult, however, because he became anxious whenever he had to fly using instruments alone. He had felt this way since the time his instruments had failed during the thunderstorm in Vietnam. His continuing insomnia and his preoccupation with thoughts about his combat experiences also interfered with his work, and he began drinking heavily in order to sleep. He was bitter that when he indicated to his commanding officer that he needed psychiatric help, he was told that "pilots don't see shrinks."

Warren's wife and children joined him at the air base. From the time he first came home, however, he felt unable to care for them or be with them. He spent his free time much as he had in Vietnam, getting sexually involved as often as possible and made several women pregnant. At times various women, or their husbands, reported his behavior to his wife. He soon sent his family back north.

On the base, Warren had an affair with a neighbor's wife who became pregnant. She wanted to leave her husband to marry him, but he told her he was leaving her and was involved with someone else. They fought, he hit her, and she formally charged him with rape. The charge was later reduced to attempted rape, but Warren remained in jail awaiting trial. After the woman admitted during the trial that she had been involved with him for some time before the alleged incident and had given him a key to her apartment, he was acquitted.

During the same period Warren cashed a number of bad checks on the base. The military authorities persuaded him to sign an agreement to leave the service in exchange for dropping charges. He said he did not realize such a discharge made him ineligible for VA benefits.

Warren returned to his family, but was at home intermittently over the next eight years, spending most of his time on flying jobs for small airline companies. The jobs involved spraying fertilizer in remote areas of the Northwest and Canada, a way for Warren to be away from people.

During this period Warren was increasingly disturbed by symptoms of posttraumatic stress. Both his insomnia and intrusive thoughts intensified during the fall and the spring, when the worst of his combat experiences had taken place. The intrusive thoughts were precursors of more disturbing episodes in which he would relive his combat experiences. Often he would become so swept up in reliving experiences that he would respond as if they were actually happening. While driving, Warren frequently reexperienced having his plane tossed back and forth in the thunderstorm. Thinking he was actually flying, he drove his car off the road on many occasions. At other times he would relive the experience of being shot by the Vietcong woman, feel the impact of the bullets, and fall backward.

These experiences left Warren fearful for his sanity and enraged over his lack of control. When the reliving experiences were frequent, he had episodes when he seemed to dissociate from his present life. On at least two occasions he found himself with women in other cities and could not recall how he had gotten there.

He also had recurrent nightmares recapitulating his combat experiences. He often dreamed of the severed head he had seen on the ground. In his dreams, however, the head would be larger than life-size and he could hear it screaming.

Warren felt guilty over his participation in missions in which American forces had landed in strange villages, frightening the people, shooting those who ran away, and interrogating those who remained. Although he had not done any of the shooting and had not been part of the interrogations, he could imagine how he would have felt had he been one of the villagers. He also felt guilty that the day he had been sick, the friend who flew in his place had been killed.

Although he had been mild-mannered before the service, a fact his family confirmed, since his return Warren had had an explosive temper. At home he would yell at his family and at times would hit his wife. He got involved in frequent fights, once breaking a bottle over the head of a man who danced with his date. On another occasion, he shot a friend in the leg who had stabbed him during a fight.

In 1969, Warren began using small amounts of heroin to stop the combat nightmares that prevented him from sleeping. Over the next few years, as greater amounts were required to control his worsening symptoms, he became addicted and began snorting heroin on a daily basis. Periodically, Warren sought help for his increasing difficulties; however, a civilian psychiatrist recommended inpatient treatment which Warren feared, and the VA repeatedly turned him away as not eligible for veterans services.

In the early 1970s, in a mood of anger and bitterness, Warren committed five bank robberies. He said he could not recall the first robbery and realized he had committed it only after waking up with his gun and $10,000 on the bed. He remembered only isolated details about the other robberies.

In 1976, he met a young prostitute in a bar. A few days later, wearing ski masks, he and the woman robbed a bank that he had planned to rob with other veterans. After this, he left his wife, began living with the woman in a common-law relationship, and had two children with her.

In the following few years, Warren was able to achieve greater stability in his life. He stopped flying in

1977 and shortly thereafter got off heroin. He began doing carpentry work on contract, and spent most of his time at home with his two small daughters. For the first time since returning from Vietnam, he began to feel a sense of love for his children. He began to recognize that when caught up in reliving combat events, what was happening was "only taking place in my head." He learned to predict when reliving experiences were about to develop and would protect himself by staying in his room. In time, he stopped acting out the experiences.

Problems developed with his second wife, however, as she tired of Warren's reclusiveness and began going out with other men. In the fall of 1980, during one of the periods when Warren's stress symptoms usually became worse, he got into an argument with her about her behavior and became abusive. She responded by getting a court order to keep him out of the house.

After he moved out, Warren felt his wife was neglecting their children, leaving them with her parents who would then leave them with a babysitter. He threatened that unless she stayed home more with the children, he would turn her and himself over to the police for the bank robbery they had committed. Instead his wife and his father-in-law decided to go to the police themselves, claimed Warren was using the robbery as blackmail to get her back, and, in return for immunity, testified against him. His wife also reported that he had been involved in five previous bank robberies before he met her and although the police had confirming films from the banks, the statute of limitations protected him from being charged with these robberies.

After he was convicted and sent to prison, Warren's insomnia, nightmares, and reliving experiences intensified and ultimately led the prison staff to refer him to us for evaluation and treatment. Warren was eager for help, and perhaps also felt that being involved in a treatment program might strengthen his chances of an earlier parole. When we initially saw him his suffering was acute, his incapacity great, and his chances for rehabilitation without successful treatment seemed slight.

Warren's stress disorder appeared to be centered on the overwhelming fear he had managed to repress in combat, and that he had subsequently expressed in his nightmares and reliving experiences. Depression and extended mourning were also a major part of his disorder. He was absorbed with the death of friends in combat, Vietnamese civilians, enemy soldiers, and his own sense of having died. Poems he wrote during and after his tour reflected this mood:

> Bodies entangled in "heaps"
> Children—"Crispy crittered"
> Mutilated
> Dead
> Women; young and beautiful
> Old and precious—for their giving,
> now cold
> broken
> fly-covered
> Even dead men
> Ours and theirs—reside inside,
> rotting in my
> head.

Of comparable significance was the bitterness he felt toward the Army for making it impossible for him to obtain treatment as a veteran. His bitterness had been expressed in behavior that was not only antisocial and destructive, but self-destructive as well. He seemed to be trying to use his anger more constructively—to get his Army discharge changed, to get benefits for his disability, to challenge his conviction and get paroled, and to get treatment.

Early in his therapy Warren had the dream of the severed head he had seen in Vietnam, and as usual it was larger than life and was screaming. He wondered why he had that dream so often, and was asked to go over the incident. On that particular flight he had been ordered to take back dead bodies. The man's body had been put on Warren's helicopter without the head Warren had seen near it, so he had told the soldiers on the ground to get the head as well. He said he could picture the man screaming when he was hit. He was asked if he had had the thought that he might wind up on the ground with his head cut off like that man. He responded strongly to this question, saying that was the one way he had not wanted to go. He said the head was the vulnerable part of a helicopter pilot, the part exposed to view and not protected by armor. Both women who, in separate incidents, had fired at him at close range, had been aiming at his head.

Warren went on to say that people killed by head wounds or head injuries comprised a major part of his Vietnam memories. He recalled the soldier who had died in his helicopter of a bullet wound in the head, crying, "Mommy, Billy is dying," and the lieutenant whose head had been smashed by the rotor blade. He then recalled several other incidents. In one a helicopter came down and got its rotor tangled with that of another which had just landed. Both pilots got chopped up and their heads were cut off. In another incident a soldier was loading a rocket from the front and left the battery switch in the "on" position. The rocket fired, blowing the soldier's head off. Warren recalled how the soldier had staggered around without his head a few seconds before falling. He ironically summarized his recollections of these incidents by saying, "People always seemed to be losing their heads."

Warren said that since returning from Vietnam, he had had no feelings for people. If someone was nice to him, he was nice in return, but without really caring. He seemed to be suggesting that, in a sense, he had functioned since the war as if his head was disembodied— or at least cut off from his heart.

He talked of his use of sex to deal with his anxiety and tension in Vietnam. He used to think on flights that if he would just make it through the day, he would be able to have sex that night. At night, he said, having sex seemed to be an assurance that he was still alive. Since his combat days he had been comfortable only with fleeting sexual encounters with women who made no claims on his emotions.

Warren was turned down on his request for parole. He was unable to discuss either combat or his post-traumatic stress symptoms with the parole board. When the board asked him why he had robbed a bank and he said he did not know, he was told that he had better stay in jail until he did. He was angry and said he felt like breaking out of the minimum security prison or blowing the whole place up. He decided, however, to concentrate on appealing the parole decision.

After the parole board rejection Warren had a reliving experience. In the incident he felt he was at a kind of funeral ceremony his Army company used to have for pilots who had been killed. They would gather together, the names would be read, taps would be played, and the bodies would be put on planes to be sent home. Warren said after a while too many pilots had died and they did not bother with the funerals anymore. He seemed to be indicating that the parole turndown was a sort of death for him, with no mourners but himself.

Warren talked of Joe, a friend who came from a town near his, who had been killed while Warren was home on leave for the funeral of his twin baby. When he first arrived home, he had called Joe's mother and they had a friendly conversation. A few days later another relative called back to tell him Joe had been killed on a mission. Warren said that "it didn't mean nothing" to him then since he had already begun to be unable to feel, but in relating the incident now he was tearful.

Warren was apprehensive about having to be in prison in February and March, the months of the Chu Lai assault mission in which so many men were lost. He was encouraged in therapy to go over the mission again, in the hope that doing so in advance of the period he feared would give him greater ability to deal with those traumatic events.

The day he had been sick and did not fly Warren had learned from the company doctor that their company "had caught hell." He went down to the airstrip and his friend Ed, a pilot who had just returned from Chu Lai, told him that another friend, Tommy, and his crew had been killed. Warren then volunteered to fly in the subsequent flights. He became tearful relating how on his first flight into the area, the helicopters landed in a straight line, and one in front of him was blown up by a mine. As the men got out of the helicopters, many of them triggered off more enemy mines. When he came back a second time Warren told the colonel in charge that he was going to pick his own spot to land. He landed on the brown spot where Tommy's plane had been hit, figuring the explosion had already detonated any mines that were in the area. He described seeing the twisted wreckage of Tommy's plane as he came down and his sense of guilt at having survived when Tommy had died. He was sure he would have died had he been in the first flight.

Following this discussion, Warren dreamed of the Chu Lai mission, of the plane in front of him exploding, and of the severed head. In the dream there was some kind of red glow so that he saw the head continuously, not intermittently by the light of exploding artillery shells. He was asked if in the actual incident he had seen the face close enough to describe it. He said he had and that was still the face he saw in his dreams. He remembered that the hair and eyebrows had been burned. It was a young face, one ear was somehow damaged, and the other one was all right. Warren imitated the screaming expression that the face appeared to have.

Warren then went on to reveal that his nightmares and other symptoms had begun in Vietnam. He dated them back to a time when he was the officer on duty and got a call that the helicopter which had gone down when it ran out of fuel had been found with skeletons in it, along with the dog tags of the missing men. The helicopter had been shot up, suspended in the tree canopy some feet off the ground. Warren figured that the Vietcong had heard the helicopters, realized they were in trouble because of the bad weather, and had found the downed plane quickly.

Warren's first flashback had occurred while taking this call. He began reliving the incidents of the day on which the helicopter had gone down. Warren said the blank spots in his memory had started at that time. He did not remember if he had even logged the call.

Warren recalled one of the dead men in particular, a pilot named Carl. He had not been that close to Carl until the night before the fatal mission. Most nights he and his friend Ed would go to the battalion mess hall, then go off the base to a brothel. The day before he crashed, Carl had told Warren that he had been flying for eight months and he knew he was not going to make it home. He showed Warren a picture of his wife and children, and said he knew he was never going to see them again. Warren tried to tell him that nobody knew when their time was up. Carl said that he had always been faithful to his wife, but that he wanted "to get laid one last time" before he died. So Ed and Warren took him along that night to dinner and to the brothel. Warren recalled that that was the first time he had ever seen Carl drink a lot.

Reflecting back on that night, Warren said there had been no particular reason for Carl to have had a premonition of danger. The next day's mission had turned out to be dangerous only because they were ordered out in such bad weather. But Warren said he felt awful about the fact that Carl had been a good Catholic and he had helped him to commit adultery. Moreover, Carl was somehow planning to die and Warren felt he had participated in the death plans.

He said that after the confirmation of Carl's death, he was never the same. Warren had held out the hope that somehow Carl and his copilot had survived and were being held as prisoners. Reflecting on that memory, he added that perhaps it was better that they were killed because of the way the Vietcong treated captured pilots. Carl had always said he would rather die than be taken prisoner.

Discussing this incident led Warren to speak again of his sense that he had died in Vietnam. He did not cry at the time of his friend's death and until quite recently, he had not been able to cry over anyone's death. Now he was able to cry, but only over things that had happened during the war. His inability to express sadness in his postwar life was consistent with his behaving as though he were dead. He spoke again of his need to pretend to feel and seemed relieved to be told that it was not necessary to pretend, because when he felt more alive his other feelings were likely to return.

Warren dreamed about a sad event in Vietnam that occurred during a period when he was flying supply missions for a South Vietnamese unit. The unit had been engaged in a firefight, and after several supply trips, Warren was ordered to fly in fresh South Vietnamese troops as reinforcements. He made a number of trips carrying troops in and carrying bodies out. Warren said that after a while he was taking out as many bodies as the number of troops he brought in. He recalled taking in an American advisor, a lieutenant, and on the next return trip he saw the lieutenant's red hair sticking out of a body bag. He tried to get the unit advisor to call in artillery or air support and was told they were not able to get it.

In frustration Warren called his commanding officer to protest that a major assault was being planned with only his helicopter for support. He said that the unit was not making any headway, that he had started to take a lot of hits, and that he was not going to continue. The officer promised to send some air support, including a gunship, and asked him to hang on until they came. As soon as the gunship came the enemy cleared out because they knew other planes were going to follow. After this mission, Warren was nicknamed "the undertaker," and someone painted a coffin on his plane.

In his dream Warren was flying out with the bodies.

He made many runs, but was not bringing replacement troops in, just taking dead ones out. Talking about the dream brought back the memory of how dead bodies used to stink on the helicopter. When he was still a junior pilot he noticed the senior pilot would fly low over the tree tops whenever they were carrying corpses, even though flying higher was safer. Once, when they were forced to fly high, he heard the belchings and gas releases that came out of the bodies because of the decrease in pressure, and the stink became even worse. He described an image that had remained with him of a moving pants leg on a body and a big bug crawling out of a hole in the dead man's fatigues.

As Warren recalled the events associated with his dream, he appeared to be keeping score in a game of death struggling with life. In the dream death seemed totally triumphant. The dream also seemed to be a continuation of the theme Warren had talked about of having brought back death from Vietnam.

In response to this observation, Warren said that he had sometimes wondered what had happened to the other men he had flown with in Vietnam. He had tried once to contact one person but had been unable to do so. He was supported in his interest to contact his former friends; not doing so seemed to be a way of acting as if they too had died.

It was also pointed out to him that in the actual incident he had been far more alive and less passive than what was suggested by his dream. He had not simply been an "undertaker," or the driver of a hearse; his protests had finally put a stop to the unnecessary dying.

Warren's mood lightened considerably after these sessions and at times he was cheerful and lively. Exploring his sense of having died in Vietnam seemed to have begun to make him feel more alive.

Much remained to be done, however. During a religious meeting conducted at the prison by outside volunteers, he had said that "dying is easy," a statement in keeping with an attitude he had professed in his sessions of having no fear of death. The woman coordinating the meeting had been critical of his remark. As she spoke Warren had a brief reliving experience in which he was back in Vietnam being fired at by the old woman whose gun contained nails. It was evident to him that he had taken the critical remarks as an attack, but in more discussion of the association it became clear that he had also reexperienced the terror he had felt when he had been fired upon. In this context, he further revealed that in his nightmares of the thunderstorm incident he would shake with fear, though had he done so in reality, he would have been unable to fly his plane.

Warren's attempts to deny the fear of death expressed in his reliving experiences and his dreams became more evident in subsequent months. He once mentioned that after the war he had deliberately put himself in life-threatening situations by frequently flying in bad weather. This appeared to be an attempt to master his fear of death by not caring whether he lived or died; it allowed him to feel triumphant either way.

In the most painful version of his thunderstorm dream, Warren would be fearful that he would scream, something his copilot had done but he had not. He associated screaming with a loss of control, the inability to fly, and with death—as in the screaming expression on the face of the severed head. During the actual storm it had been necessary for him to repress both the fear of death and the impulse to scream, but now it was equally necessary for him to be able to acknowledge and express these feelings.

After many months Warren's symptoms began to subside. For the first time since the war he went through a September, the month of the thunderstorm incident, without severe insomnia. He recalled only one dream of the thunderstorm during the entire month. There was a change from earlier versions of that dream and from reality. In the dream he voiced objections to his superiors about going out in such bad weather. In reality he had felt the objection and had even shared this feeling with a fellow pilot, but neither of them said anything to their commanding officers. The fact that Warren did so now in the dream appeared to reflect a growing awareness that in his current life he had more control over such situations.

Warren had had only one reliving experience during the last several months. It had not lasted long and he was able to joke about how he dealt with it. Usually he reacted to his symptoms with depression and anger; this lighter attitude appeared to reflect his sense of greater control. He began to experience recollections of combat as daydreams, which if unpleasant were brief, and were not accompanied by the heart palpitations, anxiety, and fear of loss of control that had accompanied his reliving experiences.

He began to talk more, and for the first time dream of his life apart from combat—his children, his parents, his brothers and sisters, and various women he knew. He remained uncertain, however, of his capacity to love again and to have any other than exploitative relationships. Anticipating parole from prison in the near future, Warren expressed his desire to continue treatment after his release. Any accurate determination of his ultimate rehabilitation will have to wait until his adaptation to life outside of prison is known.

Crime

Criminal behavior rooted in childhood, emerging in adolescence, and fully expressed in adult life, as in the case with Ray Thompson, is difficult to treat even without the complication of combat violence and posttraumatic stress. These cases point to the conclusion that while military service may make a positive contribution to confused or troubled young men, combat does not.

We cannot yet say, however, what the prognosis is for those young men who were not antisocial until after they developed posttraumatic stress. When, as with Warren Saunders, they had been living productive lives prior to combat, it may make their decline seem greater. At the same time, it is not unreasonable to think that their original adaptive capacity would indicate a better treatment prognosis.

In either case, the plight of the incarcerated veteran is one which has yet to be recognized on a wider social level. Although awareness is growing of the disproportionate number of Vietnam veterans serving time in our prisons, no systematic attempt has been made to identify those who are suffering from posttraumatic stress. The problem is most painfully evident in a case like Warren Saunders', whose criminal activity seems unlikely to have occurred had he not served his country well and developed a posttraumatic stress disorder in the process.

9

Suicide

A Presidential Commission report issued in the late 1970s estimated that Vietnam-era veterans had a suicide rate 23 percent higher than non-veterans of the same age-group.[1] A subsequent study reported that such veterans, while constituting less than 15 percent of those hospitalized in Veterans Administration facilities, accounted for almost 30 percent of inpatient suicides.[2] We suspect that the suicide rate would be much higher if the figures were restricted to Vietnam combat veterans, and higher still if they included only those with posttraumatic stress. Although suicide statistics on a national level among these different groups of veterans do not exist, those who have worked closely with men who saw heavy combat in Vietnam have noted how frequently suicidal behavior is a manifestation of posttraumatic stress disorder.[3] In our own work more than one-fifth of the first hundred veterans we have seen with posttraumatic stress have made suicide attempts, and more than another fifth have been preoccupied with suicide for a significant period of time since returning from Vietnam.

Guilt over combat experiences has played a significant role in the suicidal behavior or preoccupation with suicide of virtually all the Vietnam combat veterans with posttraumatic stress we have worked with. We have seen that guilt over combat experiences can be expressed in many ways, and that this guilt is not only a function of what a soldier actually did in combat, but also a reflection of his perception of himself while doing it.

The guilt of suicidal veterans is striking for what Robert Lifton has described as a "self-lacerating" quality. Rather than simply fearing or expecting punishment for transgressions in combat, these veterans perform "a perpetual killing of the self."[4] Greg Loomis, a suicidal veteran who had been an artillery spotter in Vietnam, was preoccupied with the memory of a friendly village that he and his sergeant had destroyed in a contest designed to see who could call in the best coordinates. Through his binoculars, Greg had watched with excitement as the shells landed. As the village was blown to bits he remembered seeing an old woman with betel nut stains on her teeth running in his direction. She was shaking her arms and legs, trying to get him to stop the shelling. As she was getting closer an artillery round exploded next to her, blowing her up.

Throughout his tour Greg thought he would be killed in action. The thought was comforting to him because it would enable him to avoid having his friends, family, and fiancée discover that he had become a cold-blooded killer in Vietnam. During the last two weeks of his tour, when he learned he was not going to be assigned to any more combat missions, he tried to kill himself with an overdose of drugs.

After more than a decade of posttraumatic stress, Greg's most painful recurrent nightmare expressed his intense guilt. In the dream he is captured by South Vietnamese villagers, strung on a pole like a pig carcass, and paraded around the village so that everyone can throw stones at him, hit him, spit on him, and call him bad names. Although the villagers yell at him in Vietnamese he understands what they are saying. The old woman with the betel nut-stained teeth and a little girl with a disfigured face are taunting him. The villagers hold him responsible for all the death and destruction in Vietnam.

Since coming home from the war Greg had had occasional disturbing episodes of altered consciousness. In one he was cleaning his garage when he thought he was back in Vietnam burning the homes of villagers. He poured gasoline over the floor and set the garage on fire and did not realize what he had done until it was in flames. During a recent suicide attempt he was reliving his Vietnam experiences, thinking he saw villagers covered with blood. He cut his wrists, feeling a sense of relief as the blood spurted out.

Even if not fully aware of the strength of his need for self-punishment, Greg was in touch with his guilt. As we have seen, many veterans are not. The consequences of being out of touch with guilt sometimes include suicidal behavior that is hard for even the veteran to explain or understand.

Tony Marco was a thirty-two-year-old Vietnam veteran referred for treatment by another veteran a few hours after he had overdosed on drugs and alcohol in a suicide attempt. A short, thin man with dark hair, he volunteered little information, and though friendly in manner spoke only in response to direct questions. He was married for the second time and had a five-year-old son from his first marriage who lived with him. He worked as an inspector for the highway department.

Tony grew up in a small town in the Northeast. His parents divorced when he was three years old and at his father's insistence, both Tony and his younger brother remained with him. Tony never asked his father about what had caused the divorce and never had any contact with his mother until she wrote to him when he was in the service. She gave him her address in case he wished to contact her, but he never did.

Tony's father and the children moved in with Tony's paternal grandmother, where they lived until his father remarried when Tony was ten. He was fond of his grandmother, and spoke with pleasure of the gatherings of his large family, which he still enjoyed.

He was happy when his father remarried, got along well with his stepmother, and liked having his own home. His stepmother took care of him, but there was no love or physical affection between them. She was more affectionate and more generous with his younger brother; however, Tony felt he was the closer of the two brothers to his father.

Tony had been a sociable adolescent with many friends and was involved in sports. From the time he entered high school, he did not like studying, and in his sophomore year he quit to go to Vietnam. He said his father and uncles had always talked of their experiences in World War II in a way that made it seem like the best part of their lives. His stepmother and father fought about his enlistment. They both wanted him to finish school, but his father understood how he felt, and signed the necessary papers. Tony's decision to go was strengthened when he learned that someone he knew had been killed during combat in Vietnam. It seemed to him that Vietnam was the place to be.

Tony enlisted in the Marines and after basic training and advanced infantry training he was sent to military transport school at Camp Lejeune, where he was trained as a maintenance mechanic. A man in his outfit had a tattoo of a snake on his chest and tried to persuade Tony to get the same tattoo, but Tony refused. One day while they were working the man asked Tony to throw him a rake and as a joke Tony threw it as a spear. It hit the man in the mouth, knocking out three teeth. After that the other man frequently made jokes about his missing teeth designed to make Tony feel guilty. He insisted that because Tony had scarred him for life, Tony would have to get the tattoo, so he too would be scarred. One night, after a drinking bout, Tony got the tattoo.

Tony was restless to go overseas and went AWOL with the idea that if he hitchhiked to the West Coast the Marines there might send him overseas. When he had gotten as far as Missouri he was picked up by MPs and spent a month of correctional duty in Camp Lejeune before being sent to Vietnam.

Tony was stationed in the Da Nang area from February 1970 until January 1971. He described Vietnam as having been "a kind of home" to him, but at the same time he had been upset by the "senselessness of it all" and by a lot of the things he saw people doing to themselves. He remembered a Marine who tried to kill himself, and also spoke of another man who, after getting a "Dear John" letter, tried to kill himself and was sent home in a straight jacket. Other recollections were of two men who got into a fight and shot each other over being first on a chow line, and of a black Marine who "fragged" himself and two other men, killing all three of them.

Tony was never in any firefights. His basic assignment was as a mechanic and truck driver, and much of the time he was assigned to escort officers on visits to various bases. Although he was in convoys that received mortar fire, and had friends who were killed on such assignments, he himself was never hit.

The most danger he experienced was in listening posts (LPs), which he was assigned to two or three nights a week. With another Marine, he would spend the night in the jungle near a suspected Vietcong transportation route, and radio back to the base if they spotted Vietcong. The two men were supposed to take turns sleeping, but Tony never wanted to sleep on LP duty and took amphetamines to stay awake. He said he had heard stories of men whose throats were cut after both partners had fallen asleep on LP duty. On these assignments Tony never spotted any Vietcong, but spent many hours frozen in silence after hearing what he now thinks were probably only the sounds of animals moving.

He was disturbed by an incident in which he witnessed a Vietnamese boy run over and dragged by a truck. Tony had helped bandage the boy and took him to the medical field base, but he was sure the boy eventually lost his leg.

Tony had several fights in Vietnam. Once he refused a sergeant's order and the sergeant knocked him out. Another time a friend read one of Tony's letters and, when they scuffled over it and the other man ripped the letter up, Tony punched him. Another incident was somewhat more serious. Tony was trying to help someone who was drunk and had passed out. A staff sergeant came by, thought Tony and the other man were fighting, and punched Tony in the eye. Tony drew his knife and was ready to kill the sergeant, but the fight was broken up by bystanders. Tony warned the sergeant not to go to sleep that night, but his anger cooled and he and the sergeant became friendly again.

When Tony was getting out of the service, his father and stepmother separated. After spending a few days at home with his father, he went to Florida to join a girl he had met on leave and stayed there for six months. He felt no one wanted to know about his Vietnam experiences, including his father, a decorated World War II veteran. At the same time, because of public sentiment against the war, Tony did not tell people he had been in Vietnam. His high-school class was graduating when he returned from Florida and Tony told his old friends he had been there since he left school.

For several years Tony worked as a police dispatcher and hoped to become a policeman; however, a fight he had with another man at work spoiled his chances. Then he became a road inspector for the state highway department, but was unhappy in his job because of the travel involved. He also did not like the fact that the work required him to be outside in bad weather much of the time.

Tony married several years after coming home from Vietnam, had one child, and divorced two years later. He said he and his first wife did not get along and that she was reluctant to have sex with him after their child was born. They had frequent fights and he occasionally slapped her. When they broke up, his wife could not handle their son alone, and Tony was glad to take him.

Tony met his second wife several years ago, and they lived together a year before marrying. He was happy with their relationship, and felt his wife and son were happy too.

Tony drank only occasionally, but was a heavy user of

marijuana, a habit he had begun in Vietnam in order to sleep after taking amphetamines. He smoked marijuana all day long, and claimed to be able to function well on it. He saw himself as "mellower," and easier to be with because of marijuana.

During most of his postcombat period, Tony had suffered from symptoms of a posttraumatic stress disorder. About two years after coming home, he began to have intrusive thoughts about Vietnam. Somewhat later, around the time of his first marriage, he started having nightmares of the war. In his most disturbing recurrent nightmare, he is sleeping when someone sneaks up from behind and cuts his throat.

Since his return from Vietnam, Tony had found it difficult to be with people and, while remaining outgoing on the surface, he felt withdrawn and distant from those around him. He also had less sexual desire than before Vietnam and was experiencing an increasing loss of interest in his work.

Tony also suffered from insomnia, was conscious of an exaggerated startle reaction, had problems in concentration, and was troubled by events that reminded him of Vietnam. Since 1975 he had had frequent bouts of depression and anxiety.

Tony's guardedness since his return was reflected in his need to carry a knife and to sit with his back to the wall in any public place. He was irritable and had an explosive temper that he felt would be worse if he did not smoke marijuana. He described an incident in which he and another man were kidding around at work and Tony suddenly lost control, grabbing the man by the throat and pushing him against the wall. Gradually Tony got control of himself and apologized. On another occasion, when his younger brother grabbed him, also in jest, Tony pulled his knife and told him never to touch him again. He occasionally became irritable with his second wife and son as well, one time grabbing his wife by the throat when she refused to do something he wanted.

In addition to his most recent suicide attempt, Tony had made several previous attempts, including one which had occurred only a few weeks before the overdose that led to his coming for help. Trying to explain how he became suicidal, he first said that no one cared for him or other Vietnam veterans. He did not include his wife and child in this declaration, however, having described himself as happy with his family. When questioned about this, he replied that from time to time he got the feeling that life was just too much of a hassle.

He tried to relate his current suicide attempt more specifically to his frustration at work. He said he was tired of being on the road, wanted a job in the highway department's main office, and felt his boss had been stalling about transferring him. As a significant factor in his suicide the job difficulties did not seem plausible since Tony conceded he could get an office job at the same or better pay elsewhere. And indeed within a month after we first saw him he got the position he wanted in the highway department.

In his fifth session, Tony reported a dream in which he was all alone in a white pick-up truck on a sandy beach and a dark, blurred figure the size of a man was coming out of the ocean toward him. He was unable to shut the window of his truck and woke up with a start. Tony connected the darkness and formlessness of the figure to the fear he had felt on LP duty in Vietnam. His inability to close the window, however, suggested the presence of a conflict underlying his nightmares, and possibly his suicidal behavior. In response to this observation Tony indicated that there were things he had not been able to reveal.

At this point it was possible to question Tony in greater detail about the recurrent nightmare in which his throat is cut, and to ask him directly if he had ever cut anyone's throat while he was in Vietnam. He responded that he had and proceeded to describe what had happened. He had been sexually involved with a girl from a small village near his base camp. One night when he came to her hut she was not there, so he sat down and waited for her. Suddenly, at the door he saw by moonlight the silhouette of a man with a gun. He grabbed his carbine, shot the man in the chest, and jumped toward him, pulling out his knife. At that moment he realized that the man was a South Vietnamese soldier. He cut the man's throat because he was afraid he might live long enough to identify him, dragged the body a distance from the hut, and left it by the road. Then he sneaked back to camp and threw away his bloody clothes. He never told anyone what had happened.

He was afraid that the South Vietnamese Army would blame the villagers and do something to them, particularly since Vietcong sympathizers had stoned the corpse while it lay by the road. Tony lived in terror that he would be caught. Eventually, it was concluded that the Vietcong had killed the soldier.

Tony claimed to have felt no guilt about this incident, only fear that he would be punished. He mentioned that the man had been given an impressive funeral, suggesting he had been well thought of. Tony also said he had thought about the man many times, and had wondered if he had been married and if he had any children.

Tony's insistence that he felt no guilt was challenged.

His recurrent dream suggested the need to be punished by the biblical principle of "an eye for an eye," or a throat for a throat. Suicide seemed a way of arranging his own punishment for what he had done.

In the next few sessions we went over the incident again. Tony said he used to tell himself that the man had no business being there and that it was his own fault he had gotten killed. He admitted that if the man he had shot had been an American, he would have called for help and taken the consequences. Despite his original denial of guilt, he revealed that he often would think of what he had done, especially while driving, and be overcome with revulsion.

After relating this incident, Tony's mood improved dramatically. He stopped having the nightmares of his throat being cut and was no longer depressed or suicidal. He had one more unpleasant dream in which someone was trying to pull him through a glass window that would have cut him badly, but he was able to resist and not get hurt. His guilt and the need for punishment were not fully resolved, but they seemed to be more under his control.

It was suggested to him that he tell his wife what had happened in Vietnam, not simply because not doing so put a barrier between them, but because it was a step in his dealing with the problem openly rather than in nightmares and suicide attempts. After about a month he was able to confide in his wife.

Tony gave the impression of a man who had not been at peace with himself, whose discontent probably predated Vietnam more than he was willing to acknowledge. His lack of purpose, his need to get away from home, and his finding Vietnam a more satisfactory home than the one he had left suggested damage to his self-esteem, the result of abandonment by his mother and coldness from his stepmother.

Tony's impulsiveness was reflected in the way he got into fights and went AWOL. His self-punitive behavior often arose from guilt over actions committed impulsively, such as throwing the rake that knocked out his friend's teeth, then getting a tattoo he did not want to appease him. His impulsive killing of the South Vietnamese soldier and subsequent guilt and suicidal behavior followed the same pattern. His suicide attempts also had an impulsive character, even more so since they occurred at times when his life was in most respects going well.

For Greg Loomis and Tony Marco, suicidal behavior seemed directly related to guilt over combat events, although they differed considerably in their degrees of consciousness of guilt. Both veterans had done things in Vietnam for which they could have been punished. Yet there are veterans like Bob Grant, who had not committed punishable acts, but whose guilt over combat experiences contributed to the breakup of his marriage, his drinking, and his work difficulties. For him, as for many other suicidal veterans, suicide came to represent a punishment for what were perceived as combat failures, failures that had become enmeshed with the symptoms of his posttraumatic stress disorder.

Bob Grant was a heavy-set, thirty-four-year-old man with a polite and cooperative manner. When he spoke, his face was almost expressionless, though his voice became tremulous discussing painful combat events. He had had numerous hospital admissions for suicide attempts and alcoholism. When first seen by us, in conjunction with one such admission, he was markedly depressed, and feeling hopeless about his future. Although he had been employed as an appliance repairman, he had not worked for the last several years because he had severed the tendons in his hand when, intoxicated, he fell through a plate glass window. He had been divorced for five years from his wife whom he married just before going to Vietnam.

Bob was born in a small town. His parents had separated when he was three and he had not seen his father since then. He and his mother moved in with her parents, a sister, and several brothers.

His mother worked and his grandmother, a strict, domineering woman, became responsible for raising him. When he was eight, his mother had another son. Bob accepted his mother's story that she had temporarily reconciled with his father and had become pregnant before the relationship ended permanently.

One of Bob's earliest memories involved a story his uncle told of coming home crying after a fight he lost. The uncle had looked to his father for sympathy and was hit by him instead. Bob said he learned two lessons from this: Don't lose a fight, and if you do, don't cry about it.

Bob appeared to have learned that lesson well. As an adolescent, he made a reputation as a fearless fighter in a gang that frequently fought with gangs in other neighborhoods. He spent little time at home. He and his friends often went for joy rides in stolen cars; he was caught once and sentenced to three-years' probation. At seventeen he dropped out of high school to hang out on the street corner or play pool until he enlisted in the Marines. Bob had wanted to be a Marine since childhood, considering it the toughest branch of service. He was trained as an infantry mechanic then volunteered for overseas duty because he was bored with his assignment in the States. Before going overseas he married his pregnant girlfriend.

Bob served with the Marines in Vietnam from November 1968 to December 1969. His unit was located in the area of the Da Nang Air Force Base, nicknamed "Rocket City" because of all the shelling it experienced. During his first six months, Bob was a machine gunner on convoys going to Phu Bai. His last six months involved security duty at Da Nang. His duties included perimeter patrol of the base and sweeps through the area to clear it of Vietcong. On most of the sweeps, he was a machine gunner in the lead jeep.

Bob said that he had been terrified of enemy fire in the jungle by day and equally frightened during the rocket fire at night. He usually drank at night to relieve his anxiety and to help him sleep during the rocket attacks. He was disturbed by his fear and would often volunteer to lead assaults, and for various missions and sweeps as a means of mastering his feelings. He described an embarassing incident in which he had burst into a hut, heard noises, opened fire, and realized too late that he had killed a bull and a cow.

After a while Bob became hardened to the rocket fire and would often sit and drink with a friend rather than run for the bunker. In the last month, when he was "short" (about to go home), he became fearful that he could be killed and never see his child.

In the course of duty Bob killed many North Vietnamese. His reaction at first was shock over having killed another human being, but after a while, he said, he came to hate the enemy and did not regard them as human. North Vietnamese torture and murders of captured Americans contributed to his feelings. Two episodes stayed with him: finding an American hanging from a tree with his penis in his mouth, and finding a dead GI whose face had been eaten by rats trapped in a burlap bag over his head.

On most occasions Bob had killed enemy soldiers from a distance. One of his most painful combat experiences was when he had actually seen the person he killed up close. His platoon was in a village and a group of Vietnamese came toward them. When Bob told them in Vietnamese to get away, everyone obeyed except one woman who kept coming. He yelled again, but she continued to approach. Finally, Bob shot her and the sight of her falling backward remained in his mind. She had been wired with two grenades, but the fact that she was a woman bothered him greatly.

Bob was choked with emotion discussing the death of Pete, his closest friend in Vietnam. Pete had extended his tour so that he and Bob could go home together. They shared a hut together, and one night they were awakened by a rocket attack. They grabbed their weapons and ran to the bunker. Bob realized he had forgotten his flak jacket and went back for it. As a result he was thirty yards behind when an incoming rocket landed in front of Pete. When Bob reached him, he saw that he had been burned, almost blown apart, the skin ripped from his face. Bob said he was crying without actual tears when someone grabbed his arm and got him to the bunker. By the time the attack ended the medics had already removed Pete's body.

When Bob came home he was distant from people, particularly his wife. He had considered himself sociable and popular before the war and was conscious of the change in his personality. He was aware that his inability to discuss with anyone the wartime experiences that were preoccupying him contributed to his sense of isolation.

Suicide

Since his return from Vietnam Bob had had frequent intrusive thoughts and nightmares about his combat experiences. In some dreams he would be with Pete in a convoy and would be firing his gun. In others his company was being hit with rockets and he would be frightened, or he would see Pete and other friends with their bodies mutilated. In another recurrent dream he would be shot and killed, but he would be like a spirit outside of his body looking at what happened. Most recently, he had dreamed he was in Vietnam, crippled and sitting in a wheel chair, being wheeled by a nurse from his ward at the hospital.

He had persistent insomnia, waking up many times during the night, and a severe startle reaction in response to loud noises. Since he experienced an increase in symptoms if he saw movies or TV programs about the war, he tried to avoid them. His difficulties in concentrating were so great that he could no longer read a book. His memory for events in Vietnam was as sharp as ever, but he had great difficulty recalling the recent past.

Bob felt guilty about the woman he had killed though he believed he had had no choice. He felt he had taken her from people who may have cared for her, and often thought how he would feel had it been his wife. He also felt guilty about the death of his friend Pete; as he was responsible for Pete's prolonging his tour in Vietnam, he felt that he should have died. At the same time, he sometimes felt lucky that Pete had been killed rather than himself, and this added to his guilt.

He was anxious, depressed, irritable, and explosive, on many occasions given to outbursts in which he punched walls and doors. He drank little before the war, began to drink more during his nights in Da Nang, and since coming home had become a heavy drinker. He was aware that drinking helped him to avoid nightmares about the war and that during his hospitalizations when he was not drinking these nightmares returned.

Although Bob and his wife had had a second child, he was involved with other women during this period, and drinking heavily. After his wife left him in 1975, he drank much more and made a number of suicide attempts. He lived with another woman on and off for two years, but continued to drink heavily, fought with the woman, and mourned the loss of his wife and children. When he damaged his hand and was unable to work, he became more depressed and suicidal. Once, intoxicated, he had jumped onto railroad tracks and attempted to touch the third rail.

Two years before we saw him, Bob had tried to make a new life for himself by going back into the service. He was accepted but when he showed up for induction intoxicated, he was told to come back the following day. The next day he missed the bus to the base. In a mood of frustration and despair, he impulsively jumped in front of a truck and was hospitalized.

Bob had cut himself off from his friends, his children, and everyone in his family except his mother, with whom he maintained a troubled relationship. He missed his children, but avoided seeing them because he felt his life was in such a mess he could not face them.

Bob was preoccupied with the failure he had made of his life; he felt worthless and that he deserved to die. A dream he reported when first seen dealt with the relationship of suicide, Vietnam, and his current life. His mother had died and was lying in a coffin. He picked up her hand and could feel there was still some life in it. Then she was hissing at him, and was out of the coffin chasing him, and he was running away. He tried to pick up something to hit her but could not. She was like a vampire.

Bob's own manner was so deathlike that several people in the hospital had commented that he looked like a walking corpse. His identification with his mother was further highlighted by his concern with the life in her hand in the dream, because the question of how much life there was in his own hand had been critical to him since cutting himself in the fall through the glass. Associating her hissing in the dream with a sound common in the speech of Oriental women also reflected his identification of his mother with the Vietnamese woman he had killed.

Bob had incorporated his mother's critical and punitive attitude toward him, and seemed to blame himself for her unhappiness as well as his own. He felt he had hurt his mother, his wife and children, and Pete, and his image of himself as destructive merged with his wartime experiences. In nightmares about the Vietnamese woman he had killed, she frequently turned from Oriental to Occidental.

Bob was severely depressed and had been treated without success for almost two years with virtually every combination of psychotropic medication. He had also been seen for six months in psychotherapy with no improvement. He did not want to be on a ward with other combat veterans or in any treatment that dealt with his experiences in Vietnam, yet he seemed to have a classic, untreated case of posttraumatic stress that had contributed to both his drinking and the breakup of his marriage. His guilt over what had happened in Vietnam had merged with his guilt over his failure to manage his civilian life. The sense that he had been crippled in Vietnam was pervasive.

Bob continued to check in and out of the hospital, admitting himself when his drinking became hopelessly out of control or after a suicide attempt and leaving as soon as the acute symptoms subsided. He said his life was over and that he could think only of the job he had lost and the family he used to have. He expressed the wish to be dead and seemed bent on destroying his life, if not through suicide, then in other slower self-destructive ways.

Bob Grant's guilt was intimately related to the loss of his friend and his wife. Yet he appeared to be particularly vulnerable to loss since his childhood experience of the breakup of his parents' marriage, which had led to the actual loss of his father and the psychological loss of his mother. Since guilt is often a factor in the posttraumatic stress disorders of veterans who are not suicidal, as well as those who are, and since vulnerability to loss has been shown to be a factor in suicide, it is important to consider how the two factors operate together in the suicidal behavior of veterans.

Gary Williams, a black divorced veteran of thirty-five, killed himself soon after he was hospitalized for treatment of posttraumatic stress. He had been born in Indiana to a family in which the father, a wounded Navy veteran of World War II, did not work. Gary felt no strong ties to him, but was close to his mother until the age of seven. At that time his father died from complications of his war wounds, and his mother remarried. Gary was sent to live with his grandmother, with whom he developed a close relationship. He returned to his mother at the age of ten, and again felt close to her, though not to his stepfather. He left home after graduating from high school and got a job in a factory. Lonely, he returned regularly to his hometown to continue a relationship with his childhood sweetheart. She became pregnant and they married shortly before he was inducted into the Army in late 1968.

During basic training Gary was deeply saddened by the death of his grandmother. He was assigned to Vietnam a few months later, where he served first with the infantry and later with the military police. During the first few months of his tour he witnessed the deaths of his two closest friends in firefights. Then he learned that his mother had died. He said her death shattered his world and he blamed it on himself for having caused her so much worry over his being in combat.

When he came back to Vietnam after leave to attend his mother's funeral, there was a dramatic change in his attitude and behavior. Before his mother died he had been afraid during the fighting; after her death he felt he had little to live for and no longer cared about his own life or the lives of others. He began to participate in the mutilation of dead enemy soldiers. He fired shots "in the general direction" of an officer he felt was giving him a hard time. During a mission he encountered some unarmed Vietnamese boys, and ordered them out of his way; when they did not respond, he opened fire and killed them. Gary also described a pattern of uncontrolled violence that involved raping, killing, and cutting up bodies of South Vietnamese women along the highway.

Gary was a squad leader and blamed himself for several incidents in which his men were killed. Even though in none of these incidents did he actually appear to be responsible, he was preoccupied by the thought that he could have prevented the deaths.

When he returned home at the end of his tour, Gary felt estranged from his wife, but they had two more children. He remained in the service and after a few years was assigned to Europe for a two-year tour; his family remained in the States. When he returned from the European assignment, Gary was put under the command of an officer he described as a brutal taskmaster.

Around this time, Gary began to have visions of his dead mother and grandmother, and heard voices ordering him to harm himself and others. He became convinced that his commanding officer was out to destroy him; he also had episodes of violence toward his wife and a neighbor. He was hospitalized and eventually discharged from the service.

After several years of continued difficulties at home, his wife left him for another man and Gary came east to work as a laborer. He had frequent fights with people he worked with in which he used blunt objects and knives as weapons. After losing his job, he came to the VA for help and was hospitalized again.

Gary had suffered from intrusive thoughts throughout the postcombat period in which he would envision scenes of dying comrades and mutilations, rapes, and murders in which he had participated. In particular, he thought about killing the South Vietnamese boys. Although he had had these thoughts since coming home from Vietnam, they had become more intense since his wife had left him.

Gary described dreams in which he would shoot the South Vietnamese boys and feel very guilty about his behavior. He also had a recurrent nightmare in which after killing enemy soldiers, he would turn the knife on himself. In still another recurrent dream he would see himself in combat fatigues lying dead in a coffin draped by an American flag; members of his family, including his wife and sister, were seated around the coffin, crying.

The loss of his grandmother, mother, and friends in combat contributed to Gary's uncontrolled violent behavior toward Vietnamese civilians. His guilt and his need for self-punishment by suicide were reflected in dreams of lying in his own coffin or actually stabbing himself to death. As Gary's behavior later bore out, dreaming of killing oneself is usually seen only in the most suicidal individuals.

Behavior such as Gary Williams' mutilation of corpses and rape and murder of Vietnamese women, Tony Marco's murder of a South Vietnamese soldier, and Greg Loomis' deliberate call for artillery fire on a friendly village was present in the combat history of two-thirds of the suicidal veterans we studied. Although, like Bob Grant, the remaining third did not participate in excessive brutality against either enemy or civilians in Vietnam, almost all of these men had killed someone at close range, an experience not nearly so common among the general population of Vietnam combat veterans. In addition, virtually all veterans who are suicidal experienced the death of close friends or buddies during combat, and as Bob Grant's case illustrates, they are much more prone than others to perceive themselves as somehow responsible for such deaths.

Whether reflecting the subconscious wish to die in as direct a way as Gary did or in a more disguised manner, almost all suicidal veterans with posttraumatic stress disorder show evidence of the type of precombat emotional loss seen in the cases we have discussed in this chapter. Almost half of the seriously suicidal veterans we have studied—including Tony, Bob, and Gary—experienced a longstanding sense of loss or abandonment in their relationship with their mothers; almost as many described themselves as feeling close to neither parent while growing up. In addition, the precombat histories of suicidal veterans are repeatedly and disproportionately marked by such events as separations from family, school failures, and disciplinary actions, which served to diminish their self-esteem and undermine their social integration.

The pervasive sense of guilt from killing and witnessing the deaths of others among suicidal veterans is understandable not only in the context of what they did or did not do in Vietnam, but also in terms of the meanings that their combat experiences had for them. These meanings appear to have derived from the interplay and meshing of the men's prewar experiences with those they encountered in Vietnam. Although overall suicidal veterans are a difficult group to treat, resolving the individual's need for self-punishment can only be accomplished, we have found, when he is able to come to terms with what his experiences in Vietnam have meant to him. And for this an understanding of precombat experiences and adaptation is often critical.

10

Substance Abuse

ONE of the earliest public images of returning Vietnam veterans linked this population with problems of substance abuse, particularly with the heavy use of addictive drugs. Although this image, like that of the uncontrollably violent veteran, did a serious disservice to the very large proportion of veterans who do not fit such characterizations, it did point to a problem which has afflicted a great many veterans in the years since returning from Vietnam.

Among the more than a hundred Vietnam veterans we have seen with posttraumatic stress disorders following heavy sustained combat, 85 percent have or have had a serious problem with either alcohol or drugs during the postcombat period. Almost two-thirds of these veterans are poly-drug abusers, misusing both alcohol and a wide variety of drugs.

A full three-quarters of all veterans we have seen with posttraumatic stress disorder are, or have been, alcohol abusers, and many of these can be classified as alcoholic. Although about 7 percent were regular precombat users of alcohol and another 30 percent began heavy drinking while in combat, the large majority of those with alcohol problems since Vietnam began regular, heavy use only during the postcombat period. Among those who started to drink heavily in Vietnam, eight out of ten have gone on to increase their use of alcohol in the years following combat.

Almost two-thirds of the veterans with posttraumatic stress disorder we have worked with have had problems with drugs during the postcombat period. Fewer than one out of five of these men had a history of drug use prior to military service. About half began their use of drugs in Vietnam, and more than one-quarter of the group began after they came home.

The fact that the overwhelming proportion of veterans having posttraumatic stress disorders have also had serious postcombat substance abuse problems suggests a causal relationship between the two conditions.[1] This relationship, however, has not been well understood by many of those treating substance-abusing veterans. Except for a brief reference to the problem by Ray Grinker and John Spiegel in their book based on work with World War II combat veterans,[2] substance abuse by persons suffering from posttraumatic stress was not discussed in the literature before Vietnam.

Much of the misinformation existing today about substance abuse by veterans stems from a failure to recognize the relationship between the use of drugs by servicemen in Vietnam and combat-related stress. There is a general consensus that, in comparison with past wars in which American forces participated, Vietnam was uniquely characterized by the widespread use of mind-altering substances by armed forces personnel, particularly during the last years of the war.³ The reason most commonly put forth to explain this phenomenon was the wide availability of high potency drugs of all kinds in Southeast Asia at a fraction of what they would cost in the States. During the late 1960s and early 1970s, marijuana, opium, amphetamines, barbiturates, and heroin were all within easy access of servicemen in Vietnam, and were openly sold in small packages resembling a pack of cigarettes. Even in small villages and hamlets, drug distributors, commonly women, were easily found, and in the larger towns drug distribution was often integrated with prostitution.

During the early 1970s a number of studies were undertaken to measure the extent of the drug problem in Vietnam.⁴ For the most part these focused on identifying the demographic, socioeconomic, and psychological characteristics of drug users, and found that drug use in Vietnam cut across all socioeconomic and racial groups of soldiers. Also cited was the fact that, in contrast to narcotics users in the United States, the majority of users in Vietnam did not have clear character disorders. None of these studies, however, specifically addressed the relationship between drugs and combat exposure, and virtually all focused on drug use on the larger bases where, particularly during the later years of the war, drug abuse was believed to be largely a function of boredom and a prevailing sense of purposelessness.

Motivated by a concern that returning troops would be bringing their drug use home with them, the military in Vietnam instituted a counter-offensive campaign, first against marijuana and later against heroin. By mid-1971 all Americans leaving Vietnam were screened through a urine-testing program. If narcotics use was detected, the soldier was given an opportunity to receive amnesty by voluntarily entering one of sixteen rehabilitation centers located throughout Vietnam to get rid of his habit before returning home. During the month of September 1971, more than 40 percent of those screened were found to have had recent use of narcotics. Twenty-five percent were found to be heavy amphetamine users and 5 percent were identified as drug addicted.[5]

The emphasis in the rehabilitation centers was on returning the individual to a drug-free state, rather than understanding or treating the roots of the substance abuse. In addition, probably because of the social sanctions given to heavy alcohol consumption in this country, and the traditionally accepted image of the hard-drinking soldier, the attention given to substance abuse by Americans in Vietnam did not include a focus on the heavy use of alcohol.

In order to investigate the degree to which the drug epidemic in Vietnam was being spread to the United States through returning veterans, a comprehensive follow-up study was initiated in mid-1971, under the auspices of the White House Special Action Office for Drug Abuse Prevention. The study, directed by Lee Robins of Washington University, was based on a sample of almost four hundred American armed forces personnel who left Vietnam during one selected month of 1971. Tracking these men in their first several years home, these researchers reported that after one year back in the States, fewer than 8 percent of those who had been addicted to drugs on leaving Vietnam showed signs of drug dependency.[6] After three years, they found that the veteran group showed no special inclination to heavier use of drugs when compared to a matched group of non-veterans, and if anything, showed a slightly greater ability to use narcotics without becoming addicted.[7]

The generally optimistic tone of the report issued following the study served to direct attention away from the fact Robins and her colleagues had reported that among those veterans who had given up narcotics use following their return home, a substantial proportion showed indications of alcoholism several years after Vietnam. Both the Robins study and other follow-up investigations reporting the same findings[8] attributed this to the greater availability and social acceptance of alcohol in civilian life, but did not address the question of why substance abuse was continuing in any form.

Another disquieting finding that was likewise underplayed by the interpretations made in the Robins study was that a large number of those who were continuing substance abuse were suffering from flashbacks, nightmares, and depression, and reported having trouble thinking clearly—all symptoms of what is now recognized as posttraumatic stress disorder. In the Robins study these symptoms were discussed as "post-Vietnam adjustment problems," resulting from, rather than possibly underlying, the individual's substance abuse.

Perhaps most troubling of all was the fact that neither the Robins study nor other analyses by the Veterans Administration[9] identified the segment of the veteran population which had begun heavy use of drugs or alcohol after their return home. Since so often among Vietnam veterans the worst symptoms of posttraumatic stress, with the concomitant use of drugs and alcohol, developed some time after combat, use of the Robins study's conclusions to minimize the problems of substance abuse among this group was particularly unfortunate. Statistics reported in the late 1970s showed that the number of hospitalized Vietnam veterans identified as alcoholics or problem drinkers more than doubled from 13 percent in 1970 to 31 percent in 1977.[10] In the latter year this group accounted for more than half of all those being treated by the VA for drug dependence problems, underscoring the prematurity of the optimistic conclusions of the Robins study.

Since neither the studies of drug use in Vietnam nor the follow-up analyses of Robins and others presented findings in a way that separated those who saw heavy combat from those having relatively low-risk support assignments, the similarity between the veterans we have worked with and the subjects of earlier research is impossible to determine. What is clear, however, from our own and others' observations of veterans with sustained substance abuse is the consistent, causative linkage between drug and alcohol behavior and combat-related stress.[11]

Some observers have suggested that the heavy use of drugs by combat soldiers in Vietnam contributed to the relatively low incidence of acute stress reactions observed in that war.[12] Among combat veterans we have studied—about one-third of whom used drugs of some kind during the war and another third who used alcohol on a regular basis—it appears that rather than preventing stress reactions, the use of substances allowed many veterans to function despite their symptoms and to complete their tours of duty, which may not otherwise have been possible. In any case, the combat histories of those whose heavy use of drugs or alcohol began in Vietnam make clear that something far more disturbing than boredom or availability was the driving force behind this behavior.

Several of the veterans whose histories have been described in this book provide examples of how the persistent use of substances in combat was often a sign of an incipient stress disorder. Ted Ford, whose sense of himself as a frightened, powerless victim of circumstances beyond his control was reflected in his identification with the baby he inadvertently killed when he shot the child's mother, used alcohol heavily during the latter part of his tour in order to cope with his fear. Similarly, Bill Clark began using alcohol excessively during combat in order to escape from the guilt he felt about having become a "mass murderer" and his sense that he was out of control. In both cases alcohol enabled these men to cope with their early stress symptoms and made it possible for them to function, even if in a reduced manner.

Tom Bradley, the veteran who repressed his guilt about his vicarious participation in the rape and murder of a Vietcong nurse, showed a similar pattern by turning to heavy use of marijuana during his last months in Vietnam. After returning to combat following several months' recuperation from wounds, Tom developed violent shaking and was reassigned to a supply base. Introduced to marijuana by the Vietnamese he worked with, he credited lasting out his tour to use of the drug. "I stayed up all night and would smoke with them and sleep all day and that's how I spent the last two months in Vietnam—just waiting to get in that bird and come on home."

Two other veterans, Don Gray, who felt his need to protect himself and his squad demanded that he stay awake, and Tony Marco, who feared getting his throat cut if he slept while on listening post duty, used amphetamines to stay awake and hypervigilant. Tony found it necessary to use marijuana when he did wish to sleep.

Veterans like Don Gray, Tony Marco, Bill Clark, and Tom Bradley gave no history suggesting a likelihood that they would have become drug or alcohol abusers if they had not developed a posttraumatic stress disorder. They had no family history of drug or alcohol abuse, had not been in trouble with the law as adolescents, and had no personal history of substance abuse—all factors which, among the general population, are significantly related to drug or alcohol abuse in adulthood.

Many of those who dealt with posttraumatic stress symptoms with drugs or alcohol, however, had precombat histories involving criminal activities and families with drinking and drug problems. Histories of maternal rejection were also usual.

Joe Prince, a black veteran, was representative of this group. Although he was a good student in high school and had no history of substance abuse prior to combat, his father and mother drank heavily, his mother went out with other men, and Joe considered himself to be the least favored of her six children. He felt no one cared for him and was torn between trying to win his mother's favor and feeling it was impossible to do so. In late adolescence, Joe got involved with a gang that committed burglaries and spent six months in jail.

Joe spent his first months in Vietnam with an infantry unit, living in a state of terror that he would be maimed or blinded. He sometimes fantasized a quick death for himself as an escape from his anxiety. Once during an ambush he was ordered to set up by his sergeant, the Vietnamese interpreter for the unit failed to heed Joe's warning to stay hidden in the bushes and was killed. Joe was blamed by the sergeant for the interpreter's death and began to be haunted by the memory of this event.

Other events that deeply disturbed Joe included firing at shadowy figures in the distance, only to discover he had killed two young boys, and gathering up the body parts of a comrade who had been blown up by a grenade. After being wounded during a firefight with a Vietcong patrol, Joe was given a map-plotting assignment at the base, where his terror of the constant shelling and the threat that the camp would be overrun continued.

During this period Joe began using heroin, purchasing it from a mama-san who visited the base daily. Shortly thereafter he became addicted. Joe was open about relating his need for heroin to his intense fear and guilt, saying, "Those last months I couldn't have gotten up in the morning without it. They'd have found me dead from all the stuff I was holding on to inside."

Steve Wallace, like Joe Prince, came from a family that paid little attention to him. He also felt rejected by his mother and a continuing need to please her. Although he did not have a precombat substance abuse problem, he had dropped out of school and was in trouble for street fighting before he went into the service. Steve's heavy use of addictive drugs began with treatment for injuries sustained in Vietnam. Both his posttraumatic stress symptoms and deep-seated character problems, however, played a role in his continuing addiction.

When Steve was sent to Vietnam in 1966 he had virtually no understanding that an actual war was going on there. During his first search-and-destroy mission, his platoon was ambushed and as Steve described it, "I remember how scared I got. I just couldn't believe all the noise and firing that broke loose. Some guy on my right was hit and bled to death lying right next to me." At that point, he said, he suddenly realized what Vietnam was all about.

A month later Steve had his first confirmed kills when eight Vietcong walked into an ambush his squad had set and he killed three of them from close range. He recalled feeling good that he was now a "veteran." He was troubled, however, by several incidents in which he killed people and did not know for sure whether they were Vietcong. Once his squad was tracking the enemy and came to a small village. A squad member began firing and when Steve saw people running out of a hut, he began firing too, killing an old man and a young boy. He stopped when his lieutenant approached and asked, "What are you guys firing about?" Although the officer did not reprimand the men, Steve was troubled about what he had done.

Throughout his combat tour Steve was constantly afraid he would be killed and felt he had to be perpetually vigilant for his own survival. He began having a nightmare, which later became recurrent, that his unit was overrun while he was sleeping. In the dream shooting and noise surrounded him; once he woke up firing his gun.

One of Steve's most painful memories was of an accident in which a railroad car full of South Vietnamese soldiers derailed when it was hit by a truck. Many of the soldiers were killed when they were knocked out of the car and it fell on top of them. Steve attempted to pull one man out, but saw that he had been severed in half. He tried to stop another man's bleeding with a tourniquet, but the man was bleeding from too many places and died.

Steve was involved in another accident in his tenth month in Vietnam in which he himself was seriously wounded. Sleeping in the field during a period when his base camp was being moved farther north, a dump truck backed over him, crushing his pelvis and ribs and causing severe kidney damage. He was evacuated to an Army hospital in Vietnam for treatment of his wounds, and spent the next year recuperating in various hospitals, first in Japan and then in the States.

Throughout his treatment Steve was given large doses of morphine to ease his pain, and within several months he was addicted. Toward the end of his recovery he was put on a sudden, abrupt detoxification program and had several days of agonizing pain from both his injuries and the lack of morphine. A hospital orderly, seeing his discomfort, offered him a shot of heroin. He quickly became part of a flourishing drug culture within the hospital, in which the cafeteria, in Steve's terms, was a "junkie hangout" where heroin was freely bought and sold.

Soon after his Army discharge, Steve married and had three children in close succession. He worked at a number of different jobs, but being with people made him irritable and he repeatedly got into fights. He was extremely troubled by insomnia, had recurrent nightmares of his Vietnam experiences, and began having the sense that people he had killed in Vietnam were following him around.

With heroin Steve was able to control the nightmares to the point where he could get a full night's rest. It also allowed him to calm down enough to be able to maintain a relationship with his family. He was unable to hold a steady job, however, and as his need for heroin increased, he became involved in a series of armed robberies. The first time he was caught he was given a suspended sentence, but following a second apprehension, he was sentenced to two years in jail.

Steve had participated in a number of drug treatment programs since being released from prison, and though he had had periods of several years free from heroin, he would substitute other drugs, especially cocaine and tranquilizers, in these interims. Considering these not nearly so effective for controlling his persistent insomnia and anxiety, he had always drifted back to heroin. At various times his distress led him to seek help, but talking about his war experiences always increased his anxiety and propelled him back to more drugs. Only when he was high on heroin, he said, could he find the peace and quiet he had lost in the war.

Steve's case illustrates a number of patterns we have consistently seen among the substance-abusing veterans we have evaluated and treated. The most significant of these concerns the functions that drugs and alcohol play in relation to these veterans' stress disorders, and the futility of attempting to get them to give up substance abuse without simultaneously alleviating stress symptoms. Although several times Steve was successful in kicking his heroin habit, his continual need for relief from his nightmares, insomnia, and deeply disturbing memories of his Vietnam experience immediately led him to other drugs and eventually back to heroin.

For the most part, Steve was able to regulate his use of heroin to allow him to function fairly well in his relationship with his wife and to maintain a high degree of intimacy with his children. Indeed, his wife often told him she preferred his behavior when he was high on heroin, equating the difference in his personality off and on the drug with "Dr. Jekyll and Mr. Hyde."

Despite their use of drugs to alleviate their posttraumatic stress symptoms, postcombat substance abuse in men like Joe Prince and Steve Wallace was partly a function of who they were prior to the Vietnam War. For the majority of veterans with posttraumatic stress, what happened to their substance abuse in the postcombat period seemed to be directly related to the vicissitudes of their posttraumatic stress disorders. Most veterans who used alcohol to deal with stress symptoms while in combat drank even more as their symptoms worsened in the postwar period. Bill Clark, for example, used alcohol in increasing amounts to alleviate the anxiety generated by his severe posttraumatic stress disorder. Only when he realized that his drinking was contributing to dangerous reliving experiences, in which he destructively reenacted events of his combat experiences, was he motivated to seek help for his stress disorder.

In many cases when substance abuse increased after the veteran came home, the substances used and the pattern of usage changed. Don Gray, for example, who had stopped amphetamine use, became dependent on alcohol after learning of the death of one of his squad members back in Vietnam and developing symptoms of a posttraumatic stress disorder. As Don was able to reestablish his protective, responsible role, particularly within his family, his stress symptoms diminished and his drinking problems disappeared. Tony Marco discontinued amphetamines when he got home, but markedly increased his use of marijuana to repress his memory of cutting the throat of a South Vietnamese soldier and to deal with the anxiety, anger, and insomnia of his stress disorder. Tom Bradley, on the other hand, switched from heavy use of marijuana to amphetamines; for several years he managed to suppress his awareness of the extent of his stress disorder, and to function in spite of the deep depression that accompanied his efforts to avoid recollections of Vietnam.

The motivation underlying the use of any particular substance appears to be directly related to its ability to alleviate specific posttraumatic stress symptoms. Marijuana, tranquilizers, and heroin tend to be used by veterans to reduce the rage associated with posttraumatic stress, to relieve the insomnia so invariably a part of the stress disorder, and to permit sleep without the interruption of combat-derived nightmares. Darvon and amphetamines enable many veterans to function socially and at work despite their combat-induced stress disorders. Alcohol, we have found, is used by different veterans to achieve almost all of the above purposes, though in some cases both alcohol and heroin become simply ways of not having to deal at all with the restrictions imposed on one's life by a stress disorder.

Some veterans attempt to balance two or more substances in order to achieve a maximal level of functioning. Such efforts present their own problems. Ray Thompson, the veteran discussed in chapter 8 who routinely participated in the torture of villagers, illustrates this pattern. During his last months in Vietnam, Ray described himself as having felt like a marked man because of the things he had done, dreading the idea of returning home and confronting how much he had changed. Avoiding missions whenever he could, he began consuming large amounts of the Scotch and bourbon his unit kept to entice the Vietnamese into cooperative behavior.

When Ray returned home his drinking increased. Although it allowed him to blot out the guilt and fear attached to his combat experiences, alcohol fueled Ray's anger and unleashed his violent behavior. After several years of experimenting with a variety of drugs, he discovered that by using heroin in combination with alcohol he was able to curb his anger and obtain at least temporary relief from his stress symptoms. Ray was unable, however, to regulate his heroin use on a long-term basis.

The relation of substance abuse to posttraumatic stress disorder is most dramatically illustrated by those veterans whose abuse began after they returned home and developed the disorder. Frank O'Donnell, whose need to deal with civilian life as an extension of the war was discussed in chapter 6, is representative of the 25 percent of the veterans we have worked with who, de-

spite arduous combat tours, reentered civilian life free
of significant drug or alcohol use, but later turned to
addictive use of substances in response to the development of posttraumatic stress symptoms. Frank's case
likewise illustrates how the attempt to medicate particular stress symptoms can generate the need for coun-
teractive medication, setting off a destructive cycle difficult to break.

After returning from Vietnam, Frank settled into his
relationship with his new wife, got a job, and enrolled
in college. During his first months home, he described
himself as having felt angry most of the time about
"nothing in particular" and often wanting to hit people.
He soon discovered that smoking marijuana lessened
the anxiety he felt when he was around people, especially at school, and within a short time he became a
daily user. When he became afraid that his sense of rage
was going to get out of control, he began to supplement
the marijuana with prescriptions obtained from the VA
for various tranquilizers.

Toward the end of his first year back in the States,
Frank's tension heightened and he began to experience
increasing pain from the combat injuries he had received in Vietnam. He obtained a prescription for Darvon from the VA and soon discovered that in addition to
reducing his physical pain, the medication had beneficial psychological effects. One Darvon, he said, could
take care of the pain. With two, he would feel pretty
good, and three would give him the feeling of being
"extremely alive and alert" for several hours. He found
Darvon relieved his anxiety better than marijuana,
made him more sociable, and improved his concentration at work and in his studies. By the end of 1972, less
than two years after coming home, he was taking at
least six Darvon capsules a day.

Because of the hyperactivity he often felt from the Darvon, Frank began increasing his use of marijuana to slow him down when he needed to do concentrated work at the end of the day. Within a short time he was smoking about one ounce per week. He also continued taking Librium and Valium when he felt really "hyped up," and frequently resorted to drinking large quantities of Scotch when he was unable to sleep.

Over the next five years, as Frank's posttraumatic stress disorder worsened, his Darvon habit continued, gradually increasing up to as many as thirty-six capsules a day. After being hospitalized for his first overdose of Darvon, Frank tried amphetamines as an alternative, and found that these gave him the same euphoric effect he had first gotten with Darvon. Frank said amphetamines would block Vietnam out of his mind and allow him to concentrate, and were more effective than anything he had ever taken to help him function. With amphetamines he could stay high for two days, but found that on the third day he would crash with exhaustion and then be unable to function at all. On these "down" days his nightmares would become worse and he would have to take massive quantities of Darvon and tranquilizers just to be able to get out of the house. Some days he did not succeed, and by 1978 his functioning was minimal.

In 1979 Frank overdosed again on a combination of Darvon and sedatives. He described the overdose as partially intentional. In the hospital, he said, he had thought about his children and vowed to get off Darvon for good. After leaving the hospital, he began taking massive quantities of aspirin in an attempt to replace the calming effects of Darvon, but in early 1980 he overdosed on those too.

At that point Frank got off pills entirely and turned

exclusively to a combination of alcohol and marijuana. Finally, realizing that his use of alcohol and marijuana could begin taking him down the same road as pills, he came into treatment and for the first time since coming home from Vietnam, began to confront in therapy the feelings that had driven his substance abuse. As his therapy progressed, both his drinking and marijuana use became more moderate.

Our work with Frank and many other veterans like him has shown that substance abuse in the context of posttraumatic stress is often significantly reduced simply as a function of psychotherapeutic treatment of the disorder. On the other hand, attempting to work on the disorder without directly intervening in veterans' substance dependency is difficult as long as they are actively engaged in the use of excessive amounts of drugs or alcohol. In some cases a carefully structured detoxification program may be needed before effective psychotherapy can begin.

In any case, it is clear that in working with veterans who suffer from both substance abuse and posttraumatic stress disorder, an exploration of the relationship between the two disorders and how they interact with one another is critical if either problem is to be alleviated. Whatever treatment is directed specifically toward the drug or alcohol abuse must reflect cognizance that the roots of the problem are frequently in the stress disorder.

Other Perspectives

11

What Protects Some?

GIVEN the trauma of intense combat in Vietnam it is not difficult to understand the high incidence of posttraumatic stress disorders among men who fought in that war, many of whom had no background of characterological disturbances or psychological difficulties.[1] It almost seems more challenging to attempt to explain what protected others who experienced combat from developing this type of disorder. Our past work,[2] and that of others,[3] has indicated that combat adaptation is related to the subsequent development of posttraumatic stress and the form it takes. Moreover, indications of a future stress disorder are often expressed in signs of emotional dyscontrol by veterans during their combat tour. In this context it appears worthwhile to examine the combat adaptation of veterans who did not develop posttraumatic stress.

Although large-scale surveys have established that there are Vietnam combat veterans who do not appear to be suffering from posttraumatic stress,[4] no detailed examination of the combat adaptation of these veterans has been made. We have undertaken an introductory analysis of a representative group of ten veterans selected because, in spite of having had traumatic combat experiences, they appeared to be dealing with postwar civilian life without evidence of a stress disorder.[5] Each participated in the study on a volunteer basis and completed the same five-session clinical evaluation we gave to Vietnam veterans suspected of having a posttraumatic stress disorder.[6]

Comparing the results of this analysis with what we have learned from our work with more than a hundred veterans with posttraumatic stress reveals several key differences between the combat adaptation of the two groups. The following case provides a reference point for beginning to identify those features that seem to play a significant role in protecting veterans from a posttraumatic stress disorder.

Paul Buckman was a slightly built, thirty-four-year-old veteran, married for eleven years, who owned and managed a garage in a small town. He had a calm, relaxed manner and related the events of his arduous combat tour in Vietnam as though these experiences were part of an unpleasant but distant past.

Paul grew up in the same town in which he now lived. He described his family as one that had a good time together despite financial difficulties. His father worked regularly as an auto mechanic, but with four children he had trouble making ends meet. When Paul was twelve his mother got a sales job in a nearby store. Recalling how he, his two older brothers, and younger sister pitched in to do chores around the house, he spoke with pride of their sense of mutual cooperation.

From his father, Paul learned how to repair cars, and

during his last two years of high school he had a part-time job as a mechanic. After graduation he became a full-time mechanic and became involved in a steady relationship with a young woman. He was satisfied with his life when, because he was about to be drafted, he enlisted in the service in late 1968.

During Paul's first months in Vietnam, his company had a commanding officer whose poor judgment, Paul felt, resulted in several unnecessary deaths. Paul's sense that he would not survive either under these conditions was heightened when, in the summer of 1969, his one-hundred-man company was overrun by enemy soldiers and two-thirds of the men were killed.

The company was reorganized and a new commanding officer assigned to it, much to the relief of the surviving men. The new officer, however, did not last a full day. The company received an order to reconnoiter an area that was to become a new firebase. As they landed by helicopter in the designated area, it became clear that it was a "hot LZ" when the lead helicopter was shot down and all the men in it, including the commanding officer, were killed. As the leader of the point squad in the company, Paul would normally have been in the first helicopter, but that day he was assigned to the last plane as a replacement for a sergeant who had been killed.

Shortly after this episode, while on a search-and-destroy mission, Paul and his squad set up an ambush and twelve enemy soldiers walked into it. He had been asleep and was awakened when the enemy triggered a Claymore mine. After the unit's machine gun jammed, Paul grabbed grenades from the other men and circling the area in which the enemy was trapped, threw them into it. He described this as the only time during his tour in which he had felt angry, something he was unable to explain other than by saying it may have had to do with being startled out of his sleep. Although Paul later received a Bronze Star for that action, he felt he had been reckless in the way he had exposed himself to danger. In a similar ambush two days later, he fired at the enemy but was careful not to expose himself in the same way.

Paul received a second Bronze Star for another close encounter with the enemy in which he was fired on from only a few feet away after he had thrown a grenade into a North Vietnamese bunker. Although his hand was nicked by a bullet, he got away by crawling on his knees. Paul described the "shakes" he had had for several days after that incident. In time, however, he said his fear in combat lessened as he became more experienced. In the beginning, he said, "Every time you saw something move you'd think it was the enemy." After a while he learned to be able to tell when there was danger and when there was not so that he avoided a perpetual state of tension.

One of Paul's disturbing recollections of Vietnam was an action in which women and children were unnecessarily killed. His company had been told that anyone they saw in a particular area would be unfriendly. When they came upon a group of people walking through a field, the lead platoon fired only to realize later that they had killed women and children. Paul's platoon was in the rear that day and he considered it fortunate that he was not directly involved.

Throughout his tour Paul felt badly about having to kill enemy soldiers because it seemed so pointless. As he put it, "You killed people in one place and then had to do it again in the same place or some place nearby. It wasn't that you took territory and then moved on to the next area."

Paul considered that much of what happened to anyone in Vietnam was a function of leadership and was
proud of the fact that none of the men in his squad was
killed during his tour. Part of being a good leader, he
felt, was explaining to the men what their specific mission was for each day, and he made a point of passing
along the information he received from his superiors to
the men in his squad. He thought it was demoralizing to
be "slogging away feeling it don't mean nothing."

When Paul returned from Vietnam he went back to
work as a mechanic and resumed the relationship with
his girlfriend, marrying a year later. He described his
wife as a thoughtful, caring person and from the beginning was happy and satisfied in his marriage. After several years, he bought his own garage and ran it with the
assistance of one of his brothers. After his retirement,
his father also helped out a few hours a day. Paul's
parents were in good health, and he had remained close
to them as well as to his siblings.

Although Paul had never developed a posttraumatic
stress disorder, he had some residue of his combat experiences. There had been times, soon after he returned, when he would drive around alone at night
thinking of things he had experienced during the war.
He had occasional dreams about his company being
overrun, landing in the hot LZ, and being fired at from
close range by the North Vietnamese soldiers, but these
eventually stopped. Rather than feeling "numb" or socially withdrawn, he said he had generally felt more
self-confident and talkative since returning from Vietnam, adding that his wife had also noticed this difference.

Paul had occasionally had startle reactions, but did
not suffer from insomnia, experienced no difficulties
with his concentration, and had no combat-related

guilt. Neither did he report any significant episodes of anxiety, explosiveness, or depression. Indeed, there was virtually no evidence that his combat experiences had had a lasting negative effect on his life.

In looking for an explanation of how Paul and others like him emerged intact from his war experiences one is struck by the number of fortuitous occurrences that protected them both physically and psychologically. Paul would ordinarily have been in the lead helicopter that was wiped out in the hot LZ, but escaped this fate because of a chance reassignment. He also could easily have been killed by the enemy soldier who fired at him from close range, but the bullet only nicked him.

Despite his skill as a squad leader, chance determined that no one in his squad was killed, particularly in the operation in which his company was almost decimated. Paul could also well have been in the lead platoon that mistakenly killed women and children. Had either of these experiences occurred, he would have had to deal with its potentially disturbing ramifications. Chance factors often played a role in the combat histories of the veterans who did not develop posttraumatic stress. At the same time, each man showed evidence of having brought to combat a way of thinking, feeling, and acting which appeared to constitute a more systematic form of protection.

Ability to function calmly under pressure. Veterans in this group regarded the ability to keep calm under pressure as a good soldier's most important attribute. All seemed to consider impulsiveness as a threat to individual and group survival. Paul experienced one break in his emotional control when his squad's machine gun did not fire, and he angrily took grenades from the other men, virtually single-handedly making a success of their ambush. His reaction to what he regarded as recklessness was a determination not to repeat such behavior.

Another veteran who did not develop posttraumatic stress following the war reported a comparable experience to Paul's that he also described as the only time he felt real anger in Vietnam. His platoon had been called in to help out another that had been ambushed and suffered heavy casualties. Enemy fire from a nearby bunker was preventing helicopters from evacuating the wounded, among whom the veteran recognized a close friend, and the platoon sergeant had "flipped out." In a mood of frustration and anger the veteran told everyone nearby to clear out. Standing up and grabbing hold of a large anti-tank M79 bazooka that was seldom used because of its dangerous recoil, he began firing and succeeded in wiping out the enemy behind the bunker. Although he was decorated for his actions, like Paul he said that for several days afterward he felt foolish for having put himself in such danger and admonished himself never to do anything like that again.

Another veteran in this group who had taken pride in his competence as a professional soldier, his ability to deal calmly with life-threatening situations, and the respect given to him by both the men in his squad and his superiors, volunteered for a second tour of duty in Vietnam. He was convinced that his skill as a soldier would prevent anything from happening to him. After six months of his second tour, a close friend, whom he had regarded as an equally skilled and professional soldier, was killed in action. This changed his feeling about his own sense of vulnerability, and he took the first available opportunity to get out of combat, a response which reflected his recognition that the emotional balance essential for functioning in combat might be deserting him.

Importance placed on understanding and judgment. Consistent with the emphasis the men in this group placed on emotional control as essential for proper decision-making was the importance they attributed to intellectual understanding of their missions' objectives and strategies as a means of controlling the stress experienced during combat. These men strove to find purpose in their combat actions, even when the situation appeared chaotic. At the same time they appeared to tolerate better than most of those who developed posttraumatic stress the lack of apparent purpose or structure which was so much a part of the war in Vietnam. All of these veterans evidenced an ability to deal with the limited objectives of each day's mission. Those who developed stress disorders following combat, on the other hand, tended to see the war in less manageable, less rational terms which often permitted the sense of chaos generated by the war to dominate them. Among veterans with posttraumatic stress was a widespread sense of the utter meaninglessness of the conflict and of being out of control in the face of it, a sense best expressed in the phrase frequently used by combat soldiers in Vietnam that whatever was going on "don't mean nothing."

Sometimes the exercise of judgment required these men to take responsibility for countermanding orders from their superior officers. One veteran, for example, described an incident in which his squad had set an ambush and about sixty Vietcong came by. When he realized that the Claymore mines his squad had set would kill only about half of the enemy, leaving his squad at the mercy of the other thirty Vietcong, he went against standing orders and directed his men not to activate the mines. After the Vietcong passed by, he called in to request artillery fire in the direction they appeared headed. Another veteran told of a similar incident in which his squad was instructed to do a body count at night in an unsafe area. Considering it foolish to go out before daybreak, he lied and told his officer the count had been completed.

In no case did such disobedience appear to express defiance or a need to simply challenge authority. Rather, these men trusted their own values and judgment and made choices consistent with both effectiveness and survival.

Acceptance of fear in self and others. Paul had "the shakes" for a few days after he was fired at from point blank range. He accepted his fear as an appropriate reaction to what had happened, realized that it would subside, and was not ashamed of what he felt. He shared with other veterans in the group the feeling that experience increased the ability to distinguish danger.

As a group those who did not develop postcombat stress disorders were also accepting of signs of fear in their comrades. Rather than condemning the expression of fear in others, they tended to explain it as due to a man's inexperience or a momentary lapse. This was reflected in the episode one veteran described of witnessing a comrade shoot and kill a young Vietcong boy who was swimming in a bomb crater filled with water. Recognizing the youth was not a threat since his gun was lying some yards away, this veteran felt the boy should have been taken prisoner and reacted with sorrow to his death. Yet he did not condemn the soldier who killed him, saying, "That guy didn't act out of a feeling of hatred, nor was he into killing. He was just frightened." This reaction was in sharp contrast to those of a significant number of veterans with post-traumatic stress disorders who denied or felt humiliated by their own fear in combat and were prone to condemn what they saw as cowardice in others.

Lack of excessive violence. Virtually all combat soldiers reported some excitement and sense of triumph during engagements in which they killed the enemy. Some developed a lust for killing and for more engagement. Among veterans with posttraumatic stress disorders are a high percentage who became turned on by violence and were driven by hatred of the enemy; however, we have not found this to be true of any of those who did not develop the disorder.

Although fear and guilt are emotions that most combat soldiers felt social pressure to repress, there was a more ambivalent attitude among the military in Vietnam toward rage and violence. Many considered rage to be an asset in making men better soldiers. In our experience, while excessive anger may occasionally have led some to perform heroic acts, when sustained it was more likely to lead to behavior that was self-destructive both to the individual and to his comrades. As a group the veterans who did not develop posttraumatic stress disorders tended to feel that rage and violence clouded judgment and led to dangerous mistakes.

One of the veterans in this group told of a man in his squad who would spend hours sharpening his long hunting knife and describing how he was planning to sneak up on unsuspecting Vietcong and slit their throats. Feeling this man was a danger to the unit, the veteran arranged with his squad leader to get him transferred.

Each of the veterans in this group was able to resist expressing frustration with enraged behavior toward the Vietnamese, even in cases where others in their units were doing so, and they did not dehumanize the enemy in their attitudes or speech. Several related incidents that suggested not only the absence of rage and dehumanization, but also a strong sense of humanity and compassion. One man, for example, had been with another soldier when, some distance away, they spotted people carrying litters who appeared to be part of a Vietcong hospital unit. Although the two soldiers felt they could have killed at least some of the group from their concealed position behind the tree line, they decided not to fire.

Absence of guilt. Consistent with the fact that the veterans who did not develop posttraumatic stress had maintained a high degree of emotional and intellectual control during combat and were not turned on by violence, their combat histories were characterized by an absence of actions over which they felt or needed to deny guilt. Specifically, none in the group had engaged in non-military killings of civilians, prisoners, or other Americans; in sexual abuse; or in mutilation of enemy dead. Since, as we discussed earlier, such actions, and subsequent guilt, were so frequent among veterans with posttraumatic stress disorders, their absence among the men we are discussing here is significant.

We have also seen that guilt is frequently found among combat soldiers whose killing was strictly within an official military context. Among those with posttraumatic stress we have found this guilt to be related to experiencing killing as a more than usual source of excitement or expression of rage. In contrast, the veterans in the non-posttraumatic stress group tended to kill enemy soldiers with a sense of reluctance, experiencing relatively little guilt afterward.

Another notable factor seen among the veterans without posttraumatic stress was an absence of guilt over having survived while others died or over actions required in order to survive. With the exception of Paul, who did not lose any of his closest friends or squad members during his combat tour, each of the men in this group witnessed the deaths of friends in Vietnam. Yet in spite of the sorrow this often evoked, none evidenced preoccupation with these deaths or feelings of guilt. This may be related to the fact that, for this group, relationships developed before or after the war seemed to have greater primacy in their lives than those formed during combat. In contrast, among many veterans we have seen with survivor guilt, friendships developed during the war appear to have superceded other attachments in their lives.

The cluster of traits observed in veterans who did not develop posttraumatic stress disorders—calmness under pressure, intellectual control, ability to create and impose a sense of structure, acceptance of their own and others' emotions and limitations, and a lack of excessively violent or guilt-arousing behavior—comprised an adaptation uniquely suitable for the preservation of emotional stability in a context that was often unstructured and unstable. These veterans experienced combat in Vietnam as a dangerous challenge to be met effectively while trying to stay alive. In contrast to most of those who developed postwar stress disorders, they did not perceive combat as a test of their worth as men, or as an opportunity to express anger or vengeance, or as a situation in which they were powerless victims.

Unlike Paul, for whom the Vietnam experience appeared to have been an episode in the ongoing course of his life, the next veteran received injuries in combat that dictated a fundamental alteration in his life. Yet, in Vietnam, he evidenced an adaptation remarkably similar to Paul's and in his postwar life showed a similar lack of evidence of posttraumatic stress symptomatology.

A thirty-six-year-old veteran, Chris Rowan experienced as intense and painful a combat tour as anyone we have seen. He killed in hand-to-hand combat, witnessed the death agonies of a close friend, was involved in actions in which women and children were inadvertently killed, and was wounded before the injury that cost him both legs. Yet when we saw him, he was happily married, had three children, and was successful in his work as an architect.

Chris came from an upper-middle-class family that put an emphasis on education and respect for authority. When Chris was young he was close to both parents, and during adolescence he developed a particular admiration and affection for his father.

After being drafted in 1966, Chris completed Officers Candidate School, and received additional parachute and ranger training before being assigned to Vietnam as an Air Cavalry platoon leader in the An Khe area. His battalion was flown on search-and-destroy missions to various nearby locations, including the An Loa Valley, where he was wounded the first time. In May 1967 his battalion was airlifted to reinforce Marines who were pinned down at Khe Sanh near the demilitarized zone.

Through these actions, Chris had a confirmed enemy hit count of 15. One hit was a Vietcong soldier who jumped into a ditch with Chris and began striking him on the head with the butt of his rifle after the weapon had jammed. Unable to aim his own rifle, Chris was finally able to knife the soldier to death.

Chris had extensive exposure to the wounded, the dying, and the dead, and recalled in particular an officer and close friend who had been hit in the spine and was thought to have died. When the man was put in a body bag, his neck jerked uncontrollably for several minutes before all signs of life stopped. Chris was also involved in fighting in which women and children were killed, sometimes as enemy participants and sometimes inadvertently. He described the tension and fear he had felt during his experiences, but showed no disturbance as he spoke about them now.

In late June 1967, Chris was assigned to flush Vietcong troops out of caves in rocky, hilly terrain adjoining the beach of the South China Sea. His sergeant wanted to use gasoline to burn them out, but Chris rejected this suggestion. That night while he slept, the enemy counter-attacked and he was hit by a grenade that landed between his legs. His unit's position was overrun and several of Chris' men were killed. A Vietcong soldier stood over Chris and, to make sure he was dead, fired two bullets into him with an automatic rifle.

For the next several days, Chris said, he thought he was going to die. He recalled in minute detail the medical and surgical events that occurred: the immediate emergency procedures, the amputations, the hospital in Japan, and the many reconstruction operations back in the States. He had two artificial limbs, but had mastered his prostheses so well that his limp was barely noticeable. He felt he had received excellent treatment in Army hospitals and expressed no bitterness toward the Army or the government. For his service in Vietnam he was awarded the Distinguished Service Cross, the Air Medal, Bronze and Silver Stars, and two Purple Hearts.

When he came out of the service, Chris took his education seriously, completing college and continuing school for a graduate degree in architecture. In addition to his regular job, he enjoyed teaching two evenings a week. He was also active in various organizations that supported disabled veterans.

Chris showed a good-humored disrespect for many of those in authority over him at work. He fantasized revenge on those who did not treat him well, but was not openly hostile toward superiors he did not like and did not behave in ways that were self-destructive. He was contented with putting his bosses down verbally to others.

Several years after returning from Vietnam, Chris married a woman he met at school. He saw himself as lucky in his marriage, describing his wife as "understanding" and "a saint." She was patient with his "down moods," which amounted to occasional irritability when he was under pressure at work. They had three children in close succession; the eldest, a daughter, was born with a congenital heart defect. When asked, Chris said he had no knowledge as to whether he was exposed to Agent Orange and did not link this with his daughter's ailment.

Chris had never developed a posttraumatic stress disorder. He did not appear to have withdrawn from people or to have been less responsive when he came out of the service. If anything, he had a more active social life and was more purposeful about his school and career aims.

Chris was asked whether he felt guilty about the men in his unit who were killed during the attack that followed his decision not to use gasoline to burn out the Vietcong. He replied that he did not dwell on this point and considered the entire incident to be "water under the bridge." In response to a question about why he had not used the gasoline that might have prevented his injury as well as the men's deaths, he replied that it would have been a "heinous way" to accomplish his objective. He added, with his characteristic humor, "Of course, what they did wasn't very gentlemanly either."

Although Chris acknowledged that at times he would think of friends who had died in Vietnam and that his eyes would tear when he heard taps on occasions like

Memorial Day, he was not preoccupied with the war, nor was he troubled by intrusive thoughts about his combat experiences. Shortly after coming home, Chris had several times had a particular nightmare in which he was being attacked by a group of Orientals with knives. He would always fight off the attack successfully, using his arms, feet, or a stick. The Orientals were not in uniform and the setting was not Vietnam.

The dream suggested that the absence of a posttraumatic stress disorder did not mean Chris was not making continual efforts to deal with his combat experiences. It is worth recalling that dreams in which traumatic events are successfully coped with are seen most frequently in those who are recovering, or have recovered from posttraumatic stress.

Chris' use of humor to deal with his situation was capsulized in a dream he remembered from the period when he was in a hospital in Japan on a ward with other amputees and some paraplegics. In the dream a helicopter was flying through the ward and the battalion commander was leaning out, screaming, "Out of bed, you lazy bastards!" The comic quality of that scene seemed to reflect Chris' attempt to master his personal tragedy with the same irony and humor he used to deal with difficulty in general. In a similar vein, he described with pleasure and considerable humor the shocked reaction of passengers and airline personnel when, as he was boarding a plane with his wife to go on vacation, his "foot fell off."

Chris believed he and other amputees went through three stages, beginning with a grandiose feeling from having survived and an accompanying sense that nothing could kill them. The second stage was characterized by a "why me" attitude and a sense of feeling sorry for oneself. In the third stage, Chris took stock of his situation, realized it could have been worse—had he lost his eyes or his hands, for example—and decided to see how well he could manage despite what had happened. Some veterans, he added, never get past the second stage. Chris expressed considerable contempt for self-pity, and was equally harsh and unsupportive of veterans who needed psychiatric help.

A dream Chris had during the course of our interviews provided a somewhat different picture. In the dream Pope John Paul was on his knees crying in the lap of an older nun, and the whole scene was being televised. He spoke of his admiration for the Pope's wit, ability to laugh at himself, and ability to deal with what had not been an easy life—all qualities with which he identified.

When it was pointed out that even this man, who seemed serene and to whom others turned for help, must at times have the need to cry, Chris emphatically agreed. Although his behavior indicated a strong need to repress his own sadness, the dream image suggested that beneath his wit, self-assuredness, and even grandiosity, lay feelings of vulnerability, loss, and dependency.

From the outset, Chris had made a good adjustment to military life. He considered his training in this country a challenging and positive experience, believed in the "moral rightness" of the United States presence in Vietnam, and expressed strong distaste for Fascist and Communist dictatorships. He was critical, however, of the fact that the United States had not fought the war with a total effort and wondered at the reasons for this. He said he would willingly fight again for a comparable cause.

He considered that as an officer he had had an adaptive advantage in combat because he understood the purpose of the military actions he engaged in. He thought it was harder for people to fight if they had no sense of where they were or what they were doing, and said that he had always tried to convey the logistics of combat to his men.

It is worth noting that although the overwhelming majority of those who saw combat in Vietnam were not officers, and there are no reliable figures on the incidence of posttraumatic stress among those who were, in our experience the disorder is common among those officers who experienced heavy combat. One such former officer, Philip Caputo, has written in *A Rumor of War* of his own experiences in Vietnam as a Marine lieutenant.[7] Caputo's account confirms our own observation that being an officer did not necessarily prevent the type of uncontrolled combat behavior that contributes to guilt-dominated posttraumatic stress disorders. Furthermore, the experience of Warren Saunders, an officer and helicopter pilot who was treated while he was in prison, demonstrates that posttraumatic stress can be the price subsequently paid by an officer who, like Chris Rowan, performed admirably in an arduous combat tour. More important than having been a lieutenant is the importance Chris placed on intelligence and judgment in combat, a quality he shared with men like Paul Buckman who were not officers.

Neither Chris nor any of the others in the group who did not develop posttraumatic stress saw in combat an opportunity to change himself or his life. They did not fit the description given by Caputo of his motives for seeking combat in Vietnam. "I joined the Marines," he wrote, "mostly because I was sick of the safe, suburban existence I had known most of my life." Caputo goes on to say that he wanted ". . . a chance to live heroically. Having known nothing but security, comfort, and peace,

I hungered for danger, challenges, and violence." He had had difficulty in school, his parents felt he was irresponsible, and in his words, "I needed to prove something, my courage, my toughness, my manhood, call it what you like."[8] Caputo's account of his murderous rage in combat and his subsequent postwar reactions, while eloquently expressing the experiences of a large number of veterans suffering from posttraumatic stress disorders, offers a sharp point of contrast to the experience and adaptation of those veterans who did not develop the disorder.

One suspects that the most significant influence on the veteran's sense of himself, the quality of his life before the war, and on his perception of combat as a duty, an opportunity to prove himself, or an escape from home, is his family. Chris and Paul, for example, grew up in supportive, caring families in which they were very close to their parents and siblings and they had remained close over the years. Five of the ten combat veterans we have seen who did not develop posttraumatic stress disorders came from warm supportive families like Paul's and Chris'. Although such a statistic in itself is not of the order to indicate any real significance, less than a handful of the hundred veterans we have seen with the disorder came from comparable families. What this suggests is that family background, in interaction with many other factors, can operate to protect veterans from stress disorders following combat.

That combat adaptation and resistance to a postcombat stress disorder is determined by more than the presence of a stable, supportive family is underlined by the fact that three of the ten veterans who did not develop posttraumatic stress came from markedly disturbed family situations. Traumatic as the combat experiences of these veterans were, their ability to survive them seemed less remarkable than their ability to have survived their traumatic childhoods.

Charles Ellis, a thirty-five-year-old, single, black veteran, was representative of this group. Coming out of a situation that seemed devoid of opportunity or hope, he had managed to achieve an education, a career, and some degree of success. The internal price he had paid was great, but his precombat adaptation seemed to have prepared him for much of what he encountered in Vietnam.

When seen by us Charles was working as the office manager for a small company. Well-dressed and neatly groomed, he was quiet and rather guarded in manner. He described himself as someone who wanted to get ahead and was willing to sacrifice to do it. He also saw himself as a private person who ordinarily did not like to discuss himself or his feelings about the war.

Charles was born out of wedlock, never knew his father, and his grandparents, who lived in another part of the country, were never told of his birth. During his early years, Charles lived with his mother and several of her relatives. Since his mother worked, one of his aunts, a deeply religious woman, looked after him. He recalled her frequently admonishing him not to fight. The aunt, the person he cared most about in his life, died when he was five. She had prepared him for her death, telling him he would have to give up his childhood, grow up in a hurry, and avoid getting into trouble. After her death, his mother's family decided that she should go back to live with her parents. Charles remained behind and for the next year was shifted from one relative to another. He had seen his mother only once since that time.

What Protects Some?

Up until the age of nine, Charles stayed primarily with one of his relatives; however, he spent much of the time in the street or at the house of a friend. When his friend's family moved south, he persuaded them to take him along, making up a story that he was going to move in with relatives in Alabama. Dropped off there, he got a job with a farmer, who also allowed him to live in his barn and to attend school. He was befriended by a teacher who noticed Charles was bright and encouraged him to pursue his education. After several years he returned to his old neighborhood in Chicago.

Charles got a part-time job working in a gas station while going to high school and began living with the station owner's family. The station owner was a gambler and taught Charles his skills. They often played as partners in high stakes card games and Charles began carrying a gun or a knife to protect himself from "sore losers." Once after winning, he was robbed at knife point. As the men who robbed him ran away he shot and wounded one of them, an incident that was never reported to the police. Charles also became a skilled pool player and supplemented his income by gambling at pool.

When Charles graduated from high school, he moved into an apartment with his girlfriend and, at the station owner's urging to make a better life for himself, got a job as a clerk in a local business. He felt he had just begun to enjoy his life when he was drafted, at the age of nineteen.

Charles served in an infantry division in Vietnam during 1967. For the entire year his outfit saw extensive combat "up and down the Delta," along the Mekong River. He received a Bronze Star and a Silver Star for his actions in combat.

The first time he came under fire Charles recalled

hitting the ground and keeping his face there the entire time. That was the only operation, however, in which he was acutely conscious of fear. Soon after that, he was part of a squad of ten men that set up an ambush. Five Vietcong walked into it and were all killed. He assumed he was responsible for some of the deaths, but there was no way of being sure.

Charles was involved twice in hand-to-hand combat. The first time several companies were brought by boat to a spot on the Delta where the enemy was known to be and after they established their positions were attacked and overrun. Charles was hit in the head with the butt of a gun, and as he turned around he shot the enemy soldier who was coming at him. He received a deep wound that required extensive suturing to close.

The second incident took place in the exact same area of the Delta shortly before the end of his tour. In that operation a friend from high school, who had been with Charles all through Vietnam, was killed. The friend was right in front of him and Charles saw him get hit. He pulled the man back to a safer area, and thinking he had not been seriously hurt, asked a medic to check him out. He was told his friend was dead.

At one point during this operation Charles was in a foxhole and enemy soldiers were throwing grenades all around the area. As one of the enemy approached him and threw a grenade, Charles grabbed him and pulled him into the foxhole, jumping out himself before the grenade went off. He assumed the man had been killed, but Charles' unit was retreating and he did not stop to look back.

A few times he almost died because he could not swim. Sometimes when outnumbered by the enemy, his company had to retreat across the Mekong River. The men who could swim well would cross the river, tie a rope to a tree, and the others would drag themselves across. The current was strong and a number of men were swept away and drowned. When Charles would get to the other side his friends would often have to pump the water out of him.

Charles described several situations in which Vietcong supporters, including women and children, threw grenades at his company and then raised their hands to surrender. He said that although they were under orders not to shoot in such circumstances, some of the men in his unit did. Although he had never participated in such actions, Charles said he could understand this reaction in men who had been injured or almost hit by the grenades.

Charles felt there was no purpose to any of the combat operations in which he participated and described fighting over and over for the same territory. He believed that the entire war was fought because the United States government wanted to get rid of his generation since "there were too many of us." Since the Vietcong always seemed to be waiting for the American troops, he was also convinced that they were informed by the American command in advance. He spoke of these things without bitterness, as though they were simply part of the situation to which one had had to adapt.

A month after being discharged from the service Charles found work selling business machines and enrolled in college. Within a short time he was promoted to field representative. After receiving his college degree some ten years later he was given a supervisory office position.

For several years after the war, Charles lived with the woman he had been involved with before going to Vietnam. They began having problems because he wanted to put his time into his schoolwork and she wanted to continue the active social life they had had when they were first together. When his studies began to suffer, he decided to leave the relationship.

After living on his own for several years, Charles met a woman in a disco. She told him that she was single and living with her sister. After several months, Charles became seriously involved with her and at that point she told him that she was married. He finally stopped seeing her because he was worried about what her husband might do if he found out. Since that relationship ended Charles had had only casual relationships with women.

In recent years his life had been severely disrupted by difficulties at work. These began when he acquired a new supervisor who Charles felt was racially prejudiced. Charles had numerous arguments with the supervisor that finally resulted in his losing his job on grounds of insubordination. Following his dismissal, Charles filed a suit against the firm because of what he considered unfair treatment.

While he was having these problems, Charles said he became depressed and paranoid, and developed high blood pressure. He also started drinking, consuming at least several beers during the evenings he spent home alone and as much as a quart of whiskey during the course of a weekend. The heavy drinking continued up until the time he was seen by us, even though he had recently obtained a better paying position in which he was well-treated.

After losing his job, Charles felt he would never be able to fully trust anyone again. He spoke of the treatment he had received from his supervisor as inflicting pain second only to what he had felt when his aunt died and he had been separated from his mother. He did not put his combat experiences in the same category.

Charles recalled that in Vietnam he had begun to wonder what he was doing there, what the meaning of it all was, and why his friend had died. Believing that he would tear himself up with such thoughts, he said he had decided to put them out of his mind and think about them later. Since the war, he had come to terms with many of these questions and, though in a sense the war was with him every day, his thoughts about it were not anxiety-provoking or intrusive. Nor was he disturbed by events that recalled the war. He had had no reliving experiences and no combat nightmares. He said, however, that he had not slept well the whole year he was in Vietnam, and had had trouble going to sleep ever since. Some nights he would lie awake for several hours, and on those nights he would often think about the war and what it all meant.

Charles was a private person before the war and remained so afterward. In his pattern of social relationships, he showed no evidence following combat of emotional numbing. At the time of his friend's death Charles had had thoughts as to whether there was anything he could have done to prevent it, but had no feelings of guilt regarding it. Nor did he feel guilty over having survived while others died.

Charles dealt with the war in a ruminative manner that seemed to work for him. He thought about it in doses he could tolerate and put aside what he could not resolve at a particular time. Although ordinarily he seemed to be a humorless man, some things that happened in combat were now sufficiently distant to appear almost comical to him. He told a story about relaxing with some other men on a patrol boat when another boat, which they had thought was friendly, began firing at them. When Charles tried to dive down on the deck for cover, he got hung up on a rail, and described in a humorous way the experience of being suspended on the rail with bullets whizzing all around him. Another time, he recalled with comic irony that after fierce firefights in which men in their unit had been killed, he and several others would risk their lives to go into town to have sex with prostitutes. These and similar stories all involved letting down his guard in some way and the attendant dangers of doing so.

Charles had been effective as a soldier and had viewed killing as necessary for survival, not as something over which he had felt out of control. When frustrated in postwar life, Charles occasionally had come close to violence, and once bought a gun with the intention of killing a man who had threatened him after losing to him in a pool game. On another occasion he took a knife up to a neighbor's apartment and threatened her after she and her children repeatedly bothered him with their noise. Rather than carry out his threats, however, he retaliated by putting speakers on his ceiling and playing loud music on his stereo during the early morning hours when his neighbors were still asleep. After that, he said, they stopped disturbing him.

When his violent feelings were aroused, Charles frequently thought back to the enemy soldier who had tried to throw a grenade at him. At such times he thought that he should have done more to the man than simply leaving him in the foxhole with the grenade. Despite his fantasies of acting on similar feelings in civilian life, he said that having survived the war without getting hurt, it did not make sense to get hurt in some kind of fight now.

What protected Charles in combat is a difficult question until one questions what had protected him while he was growing up. He seemed to have followed his aunt's injunction to grow up quickly and avoid fights and had acquired the emotional control and the practical intelligence needed for survival under trying conditions. He dealt with the pain and loss of his early life with determination to make something of himself, and by maintaining considerable emotional distance from others, he protected himself from subsequent loss.

His sense that he had managed in life with no one's help and had never needed anyone caused Charles to see himself, and to wish to have others see him, as normal and having no problems. Although, in fact, he had received very little from others, such a perception omitted the significant role played in his life by his aunt, the teacher who befriended him, and the gas station owner. While generally projecting himself as content with his current isolated lifestyle, later in his interviews he admitted he often felt lonely.

Although Charles paid a price for his early life experiences through the emotional constriction he brought to his personal relationships, he had emerged with a sense of himself and what he wanted, as well as the ability to accept people and situations for what they were. These qualities protected him in combat. The two other veterans we interviewed who came from disturbed families and who did not develop posttraumatic stress made comparable preservice adaptations. How these veterans adjusted to their early family experiences was a better predictor of their combat adaptation than the fact of the experiences themselves. These observations are consistent with the emphasis others have placed on the individual's preservice ability to tolerate anxiety as a prognostic indicator of the ability to cope with the stress of combat.[9]

The wide differences in family backgrounds among

the veterans we have seen who did not develop post-traumatic stress, together with the cases of other veterans who, despite coming from stable families, making good preservice adaptations, and functioning well in combat, did develop postcombat stress disorders, indicates that early life experience is but one of a number of factors that play a role in how well individuals adapt to the trauma of combat. It seems reasonable to assume that a caring, supportive family environment would enhance the individual's chances of developing the capacity to experience combat without attaching meanings to it derived from earlier unresolved experiences. Our work, however, points to the importance of looking more closely at perceptual and adaptive factors rather than simply at objective aspects of the precombat or combat experience to explain why some veterans have been severely distressed following their return from Vietnam while others have not. From what we have learned from veterans both with and without postcombat stress disorders, it is not what the individual experienced in his life before or during the war, but how these events and situations were perceived, integrated, and acted upon that bears the primary relationship to the postcombat response.

The fact that uncontrolled violence in combat with subsequent guilt was the pattern of a significant percentage of combat veterans who developed posttraumatic stress disorders, and not at all the pattern in our small group of those who did not develop disorders, suggests that combat behavior plays a significant role in the development of posttraumatic stress. Combat behavior in turn is related to precombat adaptation in which family stability plays a role that is not simple or decisive.

A veteran who experienced sustained, intense combat involving killing enemy soldiers at close range or witnessing the violent death of comrades is at high risk to develop a stress disorder no matter what his past history. If he was also involved in the killing of women, children, or the elderly; if he mistakenly killed civilians or other American soldiers; or if he vengefully engaged in killing of a non-military nature the likelihood of his developing the disorder appears even greater. A disturbed or delinquent precombat adaptation made many forms of guilt-arousing combat behavior more likely, and coming from an unstable family contributed to such behavior.

A veteran whose combat behavior was characterized by modulated emotional responses, absence of violent or guilt-arousing behavior, and an ability to provide his own structure in a structureless situation is less vulnerable to posttraumatic stress and is particularly less likely to develop the virulent form of the disorder characterized by guilt and self-punitive behavior. This veteran is apt to have seen military service as a duty to be performed well, but not as an opportunity to transform his life, or as an outlet for frustration and rage. An integrated, supportive family, and even more important, a stable precombat adaptation, contributed to the likelihood that combat would be dealt with in this manner.

Among veterans who were exposed to combat and who had both an unstable precombat history and a history of non-military violence in Vietnam, we have seen none who have escaped a posttraumatic stress disorder. On the other hand, we have seen veterans whose precombat histories were stable and whose combat behavior was mature, who nevertheless developed posttraumatic stress. No absolute protection is possible, but precombat and combat adaptation appear to be factors of considerable significance in determining what protects some from developing a posttraumatic stress disorder.

12

Conclusion

POSTTRAUMATIC stress is likely to be frequent among men exposed to combat in any war. Yet we have tried to convey through the cases presented in the preceding chapters the special nature of the Vietnam experience and its aftermath.

As we have seen, the meanings of combat for each individual who fought in Vietnam are singularly personal, but they are also a function of experiences that were widely shared. The shared experience in Vietnam derived largely from the unstructured and often chaotic nature of combat in that war. In contrast to previous American wars, combat in Vietnam frequently included killing of women, children, and the elderly who were sometimes the enemy, and sometimes were not. As the histories of veterans indicate, these killings were at times inadvertent, at times committed out of necessity, and at times done wantonly or vengefully. In a significant percentage of Vietnam veterans guilt contributed to the development of posttraumatic stress and to a particularly self-destructive expression of the disorder. We have seen how disabling the fear of retaliation and the need for self-punishment can be in such cases.

The psychological consequences of a war fought where the number killed, rather than the territory won was the prime objective were enormous. These consequences were multiplied by the absence of reliable guidelines for deciding who was the enemy. As a Marine captain and holder of the Navy Cross put it, "The senior officers in Vietnam appeared to have no conception of the country, the culture, or the people they were dealing with. If they ordered the evacuation of an area, they felt justified in subsequently declaring it a free-fire zone. But the Vietnamese, particularly the elderly, often knew no other way to live than to go back to their homes and their land. If they were found by young troops too caught up in war to use their own judgment, the results were disastrous." One can add that they were disastrous as well to the young men who were involved.

The young age of the average American combatant in Vietnam compared to soldiers in earlier wars was no doubt a contributing factor to their vulnerability. The absence of guilt-producing behavior was part of an ability to deal with the seemingly purposeless nature of combat in Vietnam with an inner sense of structure, purposefulness, judgment, and emotional balance. In general one would expect more such maturity in a twenty-four-year-old than in a nineteen-year-old. Although the character that one brought to combat by no means provided immunity to posttraumatic stress, it did seem to have been a significant protective factor.

Virtually every psychological disorder is influenced by the social support available to the affected individual. The mixture of terror, horror, shame, guilt, and rage that were part of combat in Vietnam made it difficult for most veterans to communicate what had happened to them, while causing family and friends to feel reluctant to share their experience. Over and over veterans have told us that their families, including fathers who were often veterans of World War II, were too uncomfortable to listen if they talked of Vietnam. And families were usually more sympathetic than outsiders, who often treated the veterans as villains for having fought or as failures for not having won.

The veteran's attempt to come to terms with his combat experience is complicated by a sense that no worthwhile purpose was served by his sacrifices and those of his comrades. The alienation of most men who fought in Vietnam from the institutions and leaders responsible for United States policy in that war is profound and contributes to their sense of living in a hostile environment. Recent scholarship on the history of the war appears to confirm the shared wisdom of the men in the field that those directing policy had little knowledge of the people they were fighting or commitment to any policy that had a chance of success.[1] And, of course, military and political policy had a direct bearing on the nature of the stress experienced by the men in the field.

As we have tried to indicate, the nature of the Vietnam experience appears to have contributed both to the development of posttraumatic stress disorders and the forms they took, but the disorder needs to be viewed and treated through an understanding of the personal meanings of combat for each individual who fought, and a knowledge of the ways in which combat events and subjective experiences were integrated by that individual. It is not enough, for example, to realize that in a particular case guilt is central to the posttraumatic stress disorder; its meaning and the defenses against

Conclusion

awareness of it will vary considerably and must be individually determined. The reciprocal relationship between guilt and fear contributes to the development of posttraumatic stress disorder, but we have seen that it also provides leverage in treatment, since any reduction in guilt is accompanied by a reduction in trauma-induced fear and vice-versa.

In earlier chapters we have shown how understanding the meaning of the combat experience is central to treatment in cases in which the postwar adaptation has a paranoid or depressive nature. The paranoid veteran must be helped toward a perception of civilian life that is not an extension of combat; the depressed veteran must be moved away from making his life a memorial to his loss. No matter what form the disorder takes, stress-oriented psychotherapy must be modified to deal with how combat-related problems have merged with problems of the veteran's postcombat life. In our experience, such treatment can be remarkably effective even with veterans whose posttraumatic stress disorder is complicated by criminal activity, suicidal behavior, or substance abuse. It appears most effective, however, in cases where these problems are clearly secondary to traumatic stress and not merely coincidental with it.

Greater understanding of posttraumatic stress on the part of mental health professionals would have a significant impact on the lives of countless veterans whose work, marital problems, substance abuse, or antisocial behavior are often seen and responded to without awareness of the underlying presence of a combat-related disorder. Tom Bradley, whose posttraumatic stress disorder centered on guilt over his passive observation of a rape and murder, had been treated in individual and marital therapy for several years with no discussion of his combat experiences or recognition of their impact on his life. His experience was common among the veterans we have seen.

Similarly, Warren Saunders, a highly decorated helicopter pilot, was sent to prison for a crime that would likely not have occurred if the posttraumatic stress disorder he had developed during the course of his arduous combat tour had been recognized and treated. There is reason to believe that a significant percentage of the incarcerated Vietnam veterans would not be in prison were it not for the lingering effects of the traumatic experiences they encountered in the course of service to their country. Certainly, the government has a responsibility for identifying and treating these men.

Veterans Administration hospitals are crowded with Vietnam veterans who are problem drinkers, dependent on drugs, or chronic alcoholics and addicts. Our work indicates that for a significant number, substance abuse is a direct result of posttraumatic stress. The dramatic cessation of drinking or drug abuse in these veterans, once they have had some relief from their stress, suggests that with proper treatment the prognosis for them is better than for those whose substance abuse is derived from other sources.

Failure to recognize posttraumatic stress is of equal consequence in civilian life, and understanding the meaning of the traumatic experience for the individual is equally helpful in treating it. One fifty-five-year-old combat veteran from World War II who suffered from nightmares and was a heavy drinker was referred to us for evaluation of a posttraumatic stress disorder. He turned out to have a stress disorder, but it was not combat-related.

At the age of forty-five this man had been happily married, had a son of thirteen to whom he was devoted, and was a partner in a successful business. While he was visiting his in-laws, he had to leave to go to work and on returning found that his wife and son had drowned in a boating accident. Two weeks after the tragedy he attempted suicide in his garage with exhaust gas, but was rescued in the nick of time by a neighbor. The man gave up his business, became a chronic alcoholic, and drifted across the country for the next ten years. Because he had told his in-laws not to let his wife and child go out in the boat alone, he blamed them for the accident and had not spoken to them for ten years.

His nightmares revolved around his anger toward his wife and son and his own guilt over what had happened. For ten years he had avoided facing or dealing with these feelings. When he began to do so in treatment his first sign of change was the resumption of his relationship with his in-laws. In a relatively short period he stopped drinking and began to make a new life for himself.

Although this man had been hospitalized several times because of his alcoholism, no attempt had been made to deal with the posttraumatic stress reaction that was the source of his drinking. Earlier recognition and treatment would probably have prevented the decade of devastation in his life.

In civilian life we are finding that stress disorders are more common than we imagined among survivors of such traumas as airplane crashes, car accidents, fires, and floods, as well as among those who, like the man described, have experienced the unexpected loss of those close to them. For the veteran, however, posttraumatic stress is the fundamental combat-derived psychological problem. This is not to say that incipient mental disorders are not precipitated by combat experience or that physical injuries sustained in combat do not cause severe psychological difficulties. But as we have seen among Vietnam combat veterans, drug and alcohol abuse, suicide, criminal or antisocial behavior, and work and marital problems often derive from stress disorders related to their combat experiences.

Posttraumatic stress appears to be the inevitable price of war for a high percentage of those exposed to sustained combat. Is there anything we can learn from the Vietnam experience that can, if not prevent, at least mitigate the effects of posttraumatic stress in those exposed to life-threatening experiences?

Our tradition has been to insist on civilian determination as to when and if men are to be used in combat, but how they are to be used and how they are to be trained have always been considered military prerogatives. Society becomes involved after a training camp disaster or an incident like My Lai, in which an American infantry company killed more than three hundred Vietnamese villagers, and public pressure requires someone to blame for the particular tragedy. There is an understandable fear on the part of the military that civilian interference could lead to a lack of discipline or toughness in recruits. The methods by which soldiers are prepared for combat, however, call for systematic reexamination on the basis of their actual effectiveness.[2]

For example, a soldier is not prepared by military training to experience his own fear in combat or the shock and horror of seeing the mutilated bodies of dying and dead comrades. Presumably military discipline is predicated on the assumption that such preparation might make men more fearful or less willing to fight. In our opinion, the avoidance of this kind of preparation is even more undermining.

One veteran told a story of seeing a training film of combat in Vietnam in which wounded South Vietnamese soldiers were being evacuated. The sergeant

Conclusion

showing the film was making the point that stupid soldiers are the ones who get hit. At that moment the commanding officer came in and ordered the film stopped
because he objected to the medical evacuation scenes.
The implication that only foolish soldiers are wounded
or killed is in itself questionable preparation for combat. The fact that the officer felt it necessary to totally
hide the reality of the situation from the recruits served
to make them far more uneasy than the film itself would
have done.

In addition, the injunction of basic training to "kill,
kill, kill," did not seem to produce better soldiers in Vietnam; it may well have created more undisciplined ones.
Given that form of training, automatic weapons that
could kill from a distance, the notion that anyone found
in a free-fire zone could be considered the enemy, and
the emphasis on body counts with few questions asked
about who had been killed, it is not surprising that
many young men became involved in unnecessary
killing.

From the standpoint of the military's own interests,
their training policy was often self-defeating. Undisciplined killing aroused guilt and fear in virtually all of
the veterans we have seen who were involved in it and
subsequently made them less effective soldiers or dangerous to their comrades. The most intractable cases of
posttraumatic stress are seen in this group after they
came home.

Undisciplined killing may raise the question of the
behavior of officers. In our experience the role of the
officers in random killing was as variable as that of the
men. Some participated in it, some condoned it, some
ignored it, and many tried to stop it, with limited success. A typical example was that of a captain who radioed a message to his men in an armored personnel car-

rier that he wished to interrogate a man they had captured. When the men came back they told him the prisoner had fallen off the carrier and had been crushed
under the tracks. He believed they were lying but he felt
it was futile to press the point. The same captain, however, was able to tell men in his company who had made
trophies of the ears of enemy soldiers that he would not
accept such behavior.

The basic manner in which the war was conducted,
from the emphasis on body counts to the declaration of
areas as free-fire zones in which anyone could be shot,
did not put the officers who saw combat—usually lieu-
tenants and seldom men of higher rank than captain—
in a position to maintain discipline. Officers who had a
second combat tour in Vietnam in the early 1970s after
tours between 1966 and 1968 confirmed the demoraliza-
tion of the troops in the last years of the war due to their
sense of the futility of their mission. The demoraliza-
tion was reflected in drug abuse, crime, and racial difficulties, which made discipline impossible. The combat
officers found themselves as much victims of political
and military policy made on higher levels as the men.

Many veterans and many who work with Vietnam
veterans are convinced that the sudden return to civilian life frequently compounded the combat soldier's
difficulties. Some men reenlisted in non-combat areas to
give themselves more time to "decompress" after their
combat experience. Time, however, was often not
enough.

Any soldier exposed to sustained combat, like any civilian exposed to traumatic stress, will benefit if he has
the opportunity to discuss his experience with a skilled
professional who will not use the information in a judg-
mental or punitive way. The discussion can be therapeutic in itself and can be used to assess the likelihood

of subsequent difficulty for the individual. Yet few soldiers returning from combat in Vietnam had such an opportunity. Had it occurred, for many veterans the warning signs of posttraumatic stress that had appeared while they were still in Vietnam—insomnia, outbreaks of rage, or episodes of uncontrollable anxiety—would have been detected.

Combat predictors of subsequent difficulty need further research, but in our experience these signs were usually present in soldiers who later developed posttraumatic stress. Had these men been helped to understand their combat and postcombat reactions, they would have been far better prepared to deal with them. As it was, many were so unprepared for the emotional after-effects of combat that their stress symptoms caused them to fear for their sanity and to withdraw into a shell of their own making.

Neither the soldier nor his family and friends back home is prepared for the full consequences of combat. Most assume that the risks of war are short-term, with the odds of survival in one's favor. If a son, husband, brother, or friend returns home from combat, those who are close to him are likely to feel relief, gratitude, and perhaps a sense of enhanced pride over his survival. Many undoubtedly expect that the growth and confidence gained from the experience will be reflected in a better future for the individual than might otherwise have been possible.

But even before Vietnam, experience had shown us that the tragic effects of combat often only begin with the end of war. The combat soldiers described by Erich Maria Remarque in *All Quiet on the Western Front*,[3] who had spent years killing and seeing comrades killed in the trenches of World War I, recognized that their lives were over whether or not they survived. The hero of the novel suggests that one man, slightly older than the rest, who had operated a farm and had a clear sense of who he was before the war, might be able to resume his life. For the rest of the group who were just beginning to find themselves, death had so permeated their lives that they were destined to remain what Remarque referred to as "a generation of men who even though they may have escaped its shells were destroyed by the war." Awareness of the human price paid by the combat soldier who has fought and survived, while not an argument against war under any circumstances, is a factor which must be included in assessing the costs of war.

Not only the individual veteran and those close to him, but the country as a whole has suffered from the lingering effects of combat on Vietnam veterans. With rare exceptions, these men came home with a sense of bitterness toward their country stemming from a feeling that their lives were dealt with carelessly by a government that had no clear sense of what it wanted from them. As one veteran put it, "In Vietnam I lost my feelings for God, for my family, and for my country. My feelings for my family and for God are returning, but I don't think I'll ever get over my anger toward the United States."

Paul Buckman, one of the veterans discussed earlier whose life seemed not to have been scarred by the war, has never voted in a national election. His refusal to vote is an expression of his belief that the war for which his friends died needlessly was waged by presidents of both parties and therefore no political leadership can be trusted. The disaffection of a large proportion of young men, most of whom went to Vietnam eager to serve their country, is one of the many sad legacies of the Vietnam War.

What will it take for the United States to get over the

wounds of Vietnam? Many who are tired of hearing about Vietnam and the Vietnam veteran predictably ask, "Isn't a memorial enough?" What the veteran needs most from his country, however, is not praise or gratitude but empathy for his experience.

The individual veteran needs to look at and come to terms with what his Vietnam experience meant to him. For the country as well, coming to terms with Vietnam and the Vietnam veteran will entail a willingness to enter into this experience. The demands of such an encounter are painful, and the urge to avoid it is great.

It took forty years for the story of the horror of the Bataan Death March to be fully told.[4] The fact that the fall of the Bataan peninsula to the Japanese in World War II was a major military defeat signaling our eventual loss of the Philippine Islands; that our men suffered unspeakable horrors as prisoners of war; that some of those horrors were inflicted by desperate, starving, and dying Americans on one another; and that almost two-thirds of the ten thousand American prisoners died, made us reluctant to explore what had happened. The soldiers who did survive, men now in their late fifties and sixties, experienced a painful satisfaction in telling their story. Reading about, or watching the television documentary based on the survivors' accounts, permits a painful but moving empathy with those men that leaves one feeling better for knowing. Vietnam veterans seem particularly moved by the story of Bataan, though many wonder why the experience had not been brought to public attention long before.

It is to be hoped that it will not take forty years for the experience of Vietnam to be shared. The recent interest in the political and military history of the war is an encouraging first sign, but an intellectual concern with the Vietnam War must include a human concern with the experience of the fighting men during and after combat. For the Vietnam experience is not simply a matter of military or political policies or mistakes. It is a devastating emotional encounter with fear, violence, guilt, death, and mourning. The country that asked its young men to fight the war that was fought in Vietnam needs to be willing to share the pain that knowledge of the experience entails.

References

1 Wounds of War

1. *New York Times,* 1 June 1971, 42.
2. Office of Research, Veterans Administration. Research on delayed stress disorders among Vietnam veterans. Washington, D.C., 1980, 1.
3. Tiffany, W. Mental health of army troops in Vietnam. *American Journal of Psychiatry* 123: 1585–86, 1967; Bourne, P.G. Military psychiatry and the Vietnam experience. *American Journal of Psychiatry* 127: 481–88, 1970; Colbach, E., and Parrish, M. Army mental health activities: 1965–1970; *Bulletin of the Menninger Clinic* 34: 333–42, 1970; Borus, J. Incidence of maladjustment in Vietnam veterans. *Archives of General Psychiatry* 30: 554–57, 1974.
4. Lifton, R.J. *Home from the war.* New York: Simon & Schuster, 1973; Egendorf, A., Kadushin, C., Laufer, R., Rothbart, G., and Sloan, L. *Legacies of Vietnam: Comparative adjustment of veterans and their peers.* Washington, D.C.: U.S. Government Printing Office, 1981; Keane, T.M., and Fairbanks, J.A. Survey analysis of combat-related stress disorders in Vietnam veterans. *American Journal of Psychiatry* 140: 348–50, 1983.
5. Egendorf, et al. *Legacies of Vietnam.*
6. Goodwin, J. The etiology of combat-related posttraumatic stress disorders. In *Posttraumatic stress disorders of the Vietnam veteran: Observations and recommendations for the psychological treatment of the veteran and his family,* T. Williams, ed. Cincinnati: Disabled American Veterans, 1980; Williams, T. A preferred model for development of interventions for psychological readjustment of Vietnam veterans. In Ibid; Wilson, J.P. Conflict, stress and growth: The effects of war on psychosocial development among Vietnam veterans. In *Strangers at home: Vietnam veterans since the war,* C.R. Figley, and S. Leventman, eds. New York: Praeger Publishers, 1980.

2 Changing Perspectives

1. Smith, J. *A review of one hundred and twenty years of the psychological literature on reactions to combat from the Civil War through the Vietnam War.* Durham, N.C.: Duke University Press, 1981.

References

2. Roscoe, T. *The web of conspiracy: The complete story of the men who murdered Abraham Lincoln.* Englewood Cliffs, N.J.: Prentice-Hall, 1959; Maxwell, J. The bizarre case of Lewis Paine. *Lincoln Herald,* 223–33, 1979.

3. Bierce, A. Ashes of the beacon. *The collected works of Ambrose Bierce,* vol. 1. New York: The Neale Publishing Co., 1909; Bierce, A. Tales of soldiers and civilians. *The collected works of Ambrose Bierce,* vol. 2. New York: The Neale Publishing Co., 1909; Sterling, G. Introduction. In Bierce, A. *In the midst of life.* New York: The Modern Library, 1927; McWilliams, C. *Ambrose Bierce: A biography.* New York: Albert and Charles Boni, 1929; Wiggins, R. *Ambrose Bierce.* Minneapolis: University of Minnesota Press, 1964; O'Connor, R. *Ambrose Bierce: A biography.* Boston: Little Brown & Co., 1967; DeCastro, A. *Portrait of Ambrose Bierce.* New York: Beekman Publishers, Inc., 1974.

4. Sterling. Introduction. *In the midst of life.*

5. Wiggins. *Ambrose Bierce.*

6. Trimble, M. *Post-traumatic neurosis: From railway spine to the whiplash.* New York: John Wiley & Sons, 1981.

7. Breuer, J., and Freud, S. *Studies in hysteria.* 2d standard edition. London: The Hogarth Press, 1955.

8. Freud, S., Ferenczi, S., Abraham, K., Simmel, E., and Jones, E. *Psychoanalysis and the war neuroses.* London: International Psychoanalytic Press, 1921, 6.

9. Ibid., 58.

10. Freud, S. *Beyond the pleasure principle.* London: Hogarth Press, 1948.

11. Kardiner, A., and Spiegel, H. *War stress and neurotic illness.* New York: Hoeber, 1947.

12. Glass, A.J. Psychotherapy in the combat zone. *American Journal of Psychiatry* 110: 725–31, 1953–54; Kolb, L.C. *Modern clinical psychiatry.* 9th ed. Philadelphia: W.B. Saunders Co., 1977.

13. Saul, L.T. Psychological factors in combat fatigue with special reference to hostility and the nightmares. *Psychosomatic Medicine* 4: 257–72, 1945; Bartemeier, L., Kubie, L., Menninger, K., Romano, J., and Whitehorn, J. Combat exhaustion. *Journal of Nervous and Mental Disease* 104: 358–89, 489–525; 1946; Trimble. *Post-traumatic neurosis.*

14. Bartemeier, et al. Combat exhaustion.

15. Futterman, S., and Pumpian-Mindlin, E. Traumatic war neuroses five years later. *American Journal of Psychiatry* 108: 401–8, 1951; Archibald, H., and Tuddenham, R. Persistent stress reaction following combat: A twenty-year followup. *Archives of General Psychiatry* 12: 475–81, 1965; Lifton, R.J. *Death in life: Survivors of Hiroshima.* New York: Random House, 1968; Niederland, W. Clinical observations on the "survivor syndrome." *International Journal of Psychoanalysis* 49: 313–15, 1968; Eitinger, L. Psychosomatic problems in concentration

References

camp survivors. *Journal of Psychosomatic Research* 13: 183–89, 1969; Lifton, R.J., and Olson, E. The human meanings of total disaster: The Buffalo Creek experience. *Psychiatry* 39: 1–18, 1976.

16. Christenson, R.M., Walker, J.I., Ross, D., and Maltbie, A. Reactivation of traumatic conflicts. *American Journal of Psychiatry* 138: 984–85, 1981; Hamilton, J.W. Unusual long-term sequelae of a traumatic war experience. *Bulletin of the Menninger Clinic* 46: 539–41, 1982.

17. Koranyi, E. Psychodynamic theories of the "survivor syndrome." *Canadian Psychiatric Association Journal* 14: 165–74, 1969; Berger, D. The survivor syndrome: A problem of nosology and treatment. *American Journal of Psychotherapy* 31: 238–51, 1977.

18. Lifton. *Death in life.*

19. Krupnick, J., and Horowitz, M. Stress response syndromes—recurrent themes. *Archives of General Psychiatry* 38: 425–35, 1981.

20. Trautman, E.C. Fear and panic in Nazi concentration camps: A biosocial evaluation of the chronic anxiety syndrome. *International Journal of Social Psychiatry* 10: 134–41, 1964.

21. Eisenhart, R.W. "You can't hack it little girl": A discussion of the covert psychological agenda of modern combat training. *Journal of Social Issues,* 31: 13–23, 1975; Shatan, C. Militarized mourning: Bogus manhood and the language of grief. Paper presented at Conference on Mythopoeisis, Santa Barbara, California, 1977.

22. Hendin, H., Pollinger, A., Singer, P., and Ulman, R. Meanings of combat and the development of posttraumatic stress disorder. *American Journal of Psychiatry* 138: 1490–93, 1981; Hendin, H., Haas, A., Singer, P., Gold, F., and Trigos, G. The influence of precombat personality on posttraumatic stress disorder. *Comprehensive Psychiatry* 24: 530–34, 1983.

23. Wilson, J.P., and Krauss, G. Predicting posttraumatic stress syndromes among Vietnam veterans. Paper presented at the 25th Neuropsychiatric Institute, Coatsville, Pennsylvania, 1982; Sudak, H., Corradi, R., Martin, R., and Gold, F. Antecedent personality factors and the post-Vietnam syndrome. *Military Medicine,* in press.

24. Danieli, Y. 1981. Differing adaptational styles in families of survivors of the Nazi Holocaust: Some implications for treatment. *Children Today* 10: 34–35. 1981.

25. American Psychiatric Association. *Diagnostic and statistical manual of mental disorders,* 3d ed. Washington, D.C., 1980.

26. Lifton. *Death in life.*

27. Holmes, T.H., and Rahe, R.H. The social readjustment rating scale. *Journal of Psychosomatic Research* 11: 213–18, 1967. Dohrenwend, B.S. Life events as stressors: A methodological inquiry. *Journal of Health and Social Behavior* 14: 167–75, 1973; Holmes, T.H., and Masuda, M. Life change and illness susceptibility. In *Stressful life*

events: Their nature and effects, B.S. Dohrenwend, and B.P. Dohrenwend, eds. New York: John Wiley & Sons, 1974.

28. Krystal, H. *Massive psychic trauma.* New York: International Universities Press, 1968.

29. Smith, J. Personal responsibility in traumatic stress reactions. *Psychiatric Annals* 12: 1021–30, 1982.

3 The Meanings of Combat

1. Saul, L.T. Psychological factors in combat fatigue with special reference to hostility and the nightmares. *Psychosomatic Medicine* 4: 257–72, 1945.

2. Lidz, T. Nightmares and the combat neurosis. *Psychiatry* 3: 37–49, 1946.

3. Kardiner, A., and Spiegel, H. *War stress and neurotic illness.* New York: Hoeber, 1947.

4. Grinker, R., and Spiegel, J.H. *Men under stress.* Philadelphia: Blakiston Co., 1945; Menninger, W. Modern concepts of war neurosis. *Bulletin of the New York Academy of Medicine* 22: 7–22, 1946.

5. Eitinger, L. Psychosomatic problems in concentration camp survivors. *Journal of Psychosomatic Research* 13: 180–83, 1969.

6. Horowitz, M. *Stress response syndromes.* New York: Jason Aronson, 1976; Lifton, R.J., and Olson, E. The human meanings of total disaster: The Buffalo Creek experience. *Psychiatry* 39: 1–18, 1976; Krupnick, J., and Horowitz, M. Stress response syndromes—recurrent themes. *Archives of General Psychiatry* 38: 428–35, 1981.

7. Hocking, F. Extreme environmental stress and its significance for psychopathology. *American Journal of Psychotherapy* 24: 4–16, 1970.

8. Shatan, C.F. Through the membrane of reality: Impacted grief and perceptual dissonance in Vietnam combat veterans. *Psychiatric Opinion* 11: 6–15, 1974; Shatan, C.F. Stress disorders among Vietnam veterans: The emotional context of combat continues. In *Stress disorders among Vietnam veterans,* C. R. Figley, ed. New York: Brunner/Mazel, 1978.

9. Haley, S.A. When the patient reports atrocities. *Archives of General Psychiatry* 30: 191–96, 1974.

10. Lifton, R.J. The postwar war. *Journal of Social Issues* 31: 181–95; 1975.

11. Wilson, J.P. Conflict, stress, and growth: The effects of war on psychosocial development among Vietnam veterans. In *Strangers at home: Vietnam veterans since the war,* C.R. Figley, and S. Leventman, eds. New York: Praeger Publishers, 1980.

References

12. Hendin, H., Pollinger, A., Singer, P., and Ulman, R. Meanings of combat and the development of posttraumatic stress disorder. *American Journal of Psychiatry* 138: 1490–93, 1981; Hendin, H. Psychotherapy for Vietnam veterans with posttraumatic stress disorders. *American Journal of Psychotherapy* 37: 86–99, 1983; Hendin, H., Haas, A., Singer, P., Gold, F., and Trigos, G. The influence of precombat personality on posttraumatic stress disorder. *Comprehensive Psychiatry* 24: 530–34, 1983.

13. Niederland, W. Psychiatric disorders among persecution victims: A contribution to the understanding of concentration camp pathology and its after-effects. *Journal of Nervous and Mental Disease* 39: 458–74, 1964.

14. Lifton, R.J. *Death in life: Survivors of Hiroshima.* New York: Random House, 1968.

15. Erikson, K. Loss of communality at Buffalo Creek. *American Journal of Psychiatry* 133: 302–5, 1976; Lifton and Olsen. The human meanings of total disaster.

16. Hendin, H., Haas, A., Singer, P., Gold, F., Ulman, R., and Trigos, G. Evaluation of posttraumatic stress in Vietnam veterans. *Journal of Psychiatric Treatment and Evaluation* 5: 303–7, 1983.

4 Stress-Oriented Psychotherapy

1. Breuer, J., and Freud, S. *Studies in hysteria.* 2d standard ed. London: Hogarth Press, 1955.

2. Freud, S. *Psychoanalysis and the war neurosis.* London: Hogarth Press, 1955; Saul, L. Psychological factors in combat fatigue with special reference to hostility and the nightmares. *Psychosomatic Medicine* 4: 257–72, 1945; Lidz, T. Nightmares and the combat neurosis. *Psychiatry* 9: 37–49, 1946; Futterman, S., and Pumpian-Mindlin, E. Traumatic war neurosis five years later. *American Journal of Psychiatry* 108: 401–8, 1951.

3. Kardiner, A., and Spiegel, H. *War stress and neurotic illness.* New York: Hoeber, 1947; Kardiner, A. Traumatic neuroses of war. In *American handbook of psychiatry,* vol. I, S. Arieti, ed. New York: Basic Books, 1959.

4. Grinker, R.P., and Spiegel, J.P. *Men under stress.* Philadelphia: Blakiston Co., 1945.

5. Horsley, J. Narcoanalysis. *Journal of Mental Science* 82: 416–22, 1936.

6. Cavenar, J.O., and Nash, J.L. Narcoanalysis: The forgotten diagnostic aid. *Military Medicine* 142: 553–55, 1977; Belson, P.M., and Dempster, C.R. Treatment of war neurosis from Vietnam. *Compre-*

hensive Psychiatry 21: 167–75, 1980; Brende, J.O., and Benedict, B. The Vietnam combat delayed stress response syndrome: Hypnotherapy of "dissociative symptoms." *American Journal of Clinical Hypnosis* 23: 34–40, 1980; Spiegel, D. Vietnam grief work using hypnosis. *American Journal of Clinical Hypnosis* 24: 33–40, 1981.

_ 7. Egendorf, A. Psychotherapy with Vietnam veterans. In *Stress disorders among Vietnam veterans: Theory, research and treatment*, C.R. Figley, ed. New York: Brunner/Mazel, 1978; Blank, A. Apocalypse terminable and interminable: Operation outreach for Vietnam veterans. *Hospital and Community Psychiatry* 33: 913–18, 1982.

8. Horowitz, M. Stress response syndromes. *Archives of General Psychiatry* 31: 768–81, 1974; Horowitz, M. *Stress response syndromes.* New York: Jason Aronson, 1976.

9. Williams, C. The mental foxhole: The Vietnam veteran's search for meaning. *American Journal of Orthopsychiatry* 53: 4–17, 1983.

10. Wilmer, H. Vietnam and madness: Dreams of schizophrenic veterans. *Journal of the American Academy of Psychoanalysis* 10: 47–65, 1982.

11. Krystal, H. Affect tolerance. *Annual of Psychoanalysis* 3: 179–219, 1975; Krystal, H. The aging survivor of the Holocaust: Integration and self-healing in posttraumatic states. *Journal of Geriatric Psychiatry* 14: 165–89, 1981.

12. Hendin, H., Haas, A., Singer, P., Houghton, W., Schwartz, M., and Wallen, V. The reliving experience in Vietnam veterans with posttraumatic stress disorder. *Comprehensive Psychiatry* 25: 165–73, 1984.

13. Krystal. The aging survivor of the Holocaust.

14. Zetzel, E. *The capacity for emotional growth.* New York: International Universities Press, 1970.

15. Krystal. Affect tolerance.

16. Kolb, L.C. The posttraumatic stress disorders of combat; A subgroup with a conditioned emotional response. *Military Medicine* 149: 237–43, 1984.

17. Marshall, J.R. The treatment of night terrors associated with the posttraumatic syndrome. *American Journal of Psychiatry* 132: 293–95, 1975; Thompson, G. Posttraumatic psychoneurosis: Evaluation of drug therapy. *Diseases of the Nervous System* 38: 617–19, 1977; Hogben, G., and Cornfield, R.B. Treatment of traumatic war neurosis with phenelzine. *Archives of General Psychiatry* 38: 440–55, 1981; Walker, J.I. Chemotherapy of traumatic war neurosis. *Military Medicine* 147: 1029–33, 1982.

18. Friedman, M. Post-Vietnam syndrome: Recognition and management. *Psychosomatics* 22: 931–43, 1981; Domash, M., and Sparr, L. Posttraumatic stress disorder masquerading as paranoid schizophrenia: Case report. *Military Medicine* 147: 772–74, 1982.

References

19. Lifton, R.J. *Home from the war.* New York: Simon & Schuster, 1973; Egendorf, A. Vietnam veteran rap groups and themes of postwar life. *Journal of Social Issues* 31: 111–24, 1975.
20. Walker, J.I., and Nash, J.L. Group therapy in the treatment of Vietnam combat veterans. *International Journal of Psychotherapy* 31: 379–89, 1981; Frick, R., and Bogart, L. Transference and countertransference as group therapy in Vietnam veterans. *Bulletin of the Menninger Clinic* 46: 429–44, 1982.
21. Brende, J.O. Combined individual and group therapy for Vietnam veterans. *International Journal of Group Psychotherapy* 31: 367–78, 1981.

5 Meanings of Guilt

1. Rado, S. *Psychoanalysis of behavior*, vol. 1. New York: Grune and Stratton, 1956; Camus, A. *The fall.* New York: Alfred A. Knopf, 1957; Rado, S. *Psychoanalysis of behavior,* vol. 2. New York: Grune and Stratton, 1962.
2. Futterman, S., and Pumpian-Mindlin, E. Traumatic war neurosis five years later. *American Journal of Psychiatry* 108: 401–8, 1951.
3. Hendin, H. Psychotherapy for Vietnam veterans with posttraumatic stress disorders. *American Journal of Psychotherapy* 37: 86–99, 1983; Hendin, H., Haas, A., Singer, P., Gold, F., and Trigos, G. The influence of precombat personality on posttraumatic stress disorder. *Comprehensive Psychiatry* 24: 530–34, 1983.
4. Rado. *Psychoanalysis of behavior,* vols. 1 and 2.

6 Combat Never Ends: The Paranoid Adaptation

1. Eisenhart, R.W. "You can't hack it little girl": A discussion of the covert psychological aspects of modern combat training. *Journal of Social Issues* 31: 13–23, 1975; Shatan, C. Stress disorders among Vietnam veterans. In *Stress disorders among Vietnam veterans: Theory, research and treatment,* C.R. Figley, ed. New York: Brunner/Mazel, 1978; Smith, C. Oral history as "therapy": Combatants' accounts of the Vietnam War. In *Strangers at home: Vietnam veterans since the war,* C.R. Figley, and S. Leventman, eds. New York: Praeger Publishers, 1980; Domash, M.D., and Sparr, L. Posttraumatic stress masquerading as paranoid schizophrenia: A case report. *Military Medicine* 147: 772–74, 1982; Lipkin, J., Blank, A., Parson, E., and Smith, J. Vietnam veter-

References

ans and posttraumatic stress disorder. *Hospital and Community Psychiatry* 33: 908–12, 1982; Brende, J.O. 1983. A psychodynamic view of character pathology in Vietnam combat veterans. *Bulletin of the Menninger Clinic* 47: 193–216, 1983; Brende, J.O., and McCann, I.L. Regressive experiences in Vietnam veterans: Their relationship to war, posttraumatic stress symptoms, and recovery. *Journal of Contemporary Psychotherapy* 14: 57–73, 1984.

2. American Psychiatric Association. *Diagnostic and statistical manual of mental disorders,* 3d ed. Washington, D.C., 1980.

3. Yager, J. Personal violence in infantry combat. *Archives of General Psychiatry* 32: 257–61, 1975; Bond, T.C. The why of fragging. *American Journal of Psychiatry* 133: 1328–30, 1976; Yager, J. Postcombat violent behavior in psychiatrically maladjusting soldiers. *Archives of General Psychiatry* 33: 1332–35, 1976.

4. Fox, R. Post-combat adaptational problems. *Comprehensive Psychiatry* 13: 435–43, 1972; Fox, R. Narcissistic rage and the problem of combat aggression. *Archives of General Psychiatry* 31: 807–11, 1974.

5. Rado, S. Psychodynamics and treatment of traumatic war neurosis (trauma to phobia). *Psychosomatic Medicine* 4: 362–68, 1942; Kardiner, A., and Spiegel, H. *War stress and neurotic illness.* New York: Hoeber. 1947.

7 Mourning Never Ends: The Depressive Adaptation

1. Niederland, W.C. The psychiatric evaluation of emotional disorders in survivors of Nazi persecutions. In *Massive psychic trauma,* H. Krystal, ed. New York: International Universities Press, 1968; Niederland, W.C. Clinical observations on the "survivor syndrome." *International Journal of Psychoanalysis* 49: 313–15, 1968; Sonnenberg, S. A special form of survivor syndrome. *Psychoanalytic Quarterly* 41: 58–62, 1972; Krystal, H. The aging survivor of the Holocaust: Integration and self-healing in posttraumatic states. *Journal of Geriatric Psychology* 14: 165–89, 1981.

2. Nace, E., Meyers, A., O'Brien, C., Ream, N., and Mintz, J. Depression in veterans two years after Vietnam. *American Journal of Psychiatry* 134: 167–70, 1977.

3. Helzer, J., Robins, L., and Davis, D. Depressive disorders in Vietnam returnees. *Journal of Nervous and Mental Diseases* 163: 177–85, 1976; Helzer, J., Robins, L., Wish, E., and Hesselbrock, M. Depression in Vietnam veterans and civilian controls. *American Journal of Psychiatry* 136: 526–29, 1979.

4. Solomon, G., Zarcone, V., Yoerg, R., Scott, N., and Maurer, R. Three psychiatric casualties from Vietnam. *Archives of General Psychiatry*

References

25: 522–24, 1971; Horowitz, M., and Solomon, G. Delayed stress response syndromes in Vietnam veterans. *Journal of Social Issues* 4: 67–80, 1975.

5. Shatan, C.F. The grief of soldiers. *American Report* 2: 1–3, 1972; Shatan, C.F. The grief of soldiers: Vietnam combat veterans self-help movement. *American Journal of Orthopsychiatry* 45: 640–53, 1973; Shatan, C.F. Through the membrane of reality: Impacted grief and perceptual dissonance in Vietnam combat veterans. *Psychiatric Opinion* 11: 6–15, 1974.

6. Jackson, H.C. Moral nihilism: Developmental arrest as a sequel to combat stress. *Adolescent Psychiatry* 10: 228–42, 1982.

8 Crime

1. May, E. Inmate veterans: Hidden casualties of a lost war. *Corrections Magazine* 5: 3–13, 1979.

2. U.S. Congress, House Committee on Veterans Affairs. *Presidential review memorandum on Vietnam-era veterans*, H.R. 38, 10 October 1978.

3. Gault, W.B. Some remarks on slaughter. *American Journal of Psychiatry* 128: 450–54, 1971.

4. Brady, D., and Rappoport, L. Violence and Vietnam: A comparison between attitudes of civilians and veterans. *Human Behavior* 26: 735–52, 1974.

5. Leventman, S. Epilogue: Social and historical perspectives on the Vietnam veteran. In *Stress disorders among Vietnam veterans: Theory, research and treatment*, C.R. Figley, ed. New York: Brunner/Mazel, 1978.

6. Bond, T. The why of fragging. *American Journal of Psychiatry* 183: 1328–30, 1976; Moskos, C. Surviving the war in Vietnam. In *Strangers at home: Vietnam veterans since the war*, C.R. Figley, and S. Leventman, eds. New York: Praeger Publishers, 1980.

7. Kardiner, A., and Spiegel, H. *War stress and neurotic illness.* New York: Hoeber, 1947.

8. Lipkin, J., Blank, A., Parson, E., and Smith, J. Vietnam veterans and posttraumatic stress disorder. *Hospital and Community Psychiatry* 33: 908–12, 1982.

9 Suicide

1. U.S. Congress, House Committee on Veterans Affairs. *Presidential review memorandum on Vietnam-era veterans*, H.R. 38, 10 October, 1978.

2. Baker, J. Monitoring of suicidal behavior among patients in the VA health care system. *Psychiatric Annals* 14: 272–75, 1984.

3. Jury, D. The forgotten warriors: New concern for the Vietnam veteran. *Behavioral Medicine* 6: 38–41, 1979; Stuen, M.R., and Solberg, K.B. The Vietnam veteran: Characteristics and needs. In *The Vietnam veteran in contemporary society.* L.J. Sherman, and E.M. Caffey, eds. Washington, D.C.: Veterans Administration, 1972; Lipkin, J., Blank, A., Parson, E., and Smith, J. Vietnam veterans and posttraumatic stress disorder. *Hospital and Community Psychiatry* 33: 908–12, 1982.

4. Lifton, R.J. *Home from the war.* New York: Simon & Schuster, 1973.

10 Substance Abuse

1. Lacoursiere, R. B., Godfrey, K., and Ruby, L. Traumatic neurosis in the etiology of alcoholism: Vietnam combat and other trauma. *American Journal of Psychiatry* 137: 966–68, 1980.

2. Grinker, R.P., and Spiegel, J.P. *Men under stress.* Philadelphia: Blakiston Co., 1945.

3. Siegel, A. The heroin crisis among United States forces in Southeast Asia. *Journal of the American Medical Association* 223: 1258–61, 1973.

4. Hampton, P., and Vogel, D. Personality characteristics of servicemen returned from Vietnam identified as heroin abusers. *American Journal of Psychiatry* 130: 1031–32, 1973; Nace, E., Meyers, A., and Rothberg, J. Addicted Vietnam veterans: A comparison of self-referred and system-referred samples. *American Journal of Psychiatry* 130: 1242–45, 1973; Nail, R., Gunderson, E., and Arthur, R. Black-white differences in social background and military drug abuse patterns. *American Journal of Psychiatry* 131: 1097–1102. 1974.

5. Zinberg, N. Heroin use in Vietnam and the United States: A contrast and a critique. *Archives of General Psychiatry* 26: 486–88, 1972; Robins, L. *The Vietnam drug user returns.* Washington, D.C.: Special Action Office Monograph, 1974.

6. Robins. Ibid; Robins, L., Davis, D., and Goodwin, D. Drug use by U.S. Army enlisted men in Vietnam: A follow-up on their return home. *American Journal of Epidemiology* 99: 235–49, 1974.

7. Robins, L., and Helzer, J. Drug use among Vietnam veterans—three years later. *Medical World News* 16: 44–49, 1975.

8. Nace, E., O'Brien, C., Mintz, J., Ream, N., and Meyers, A. Adjustment among Vietnam veteran drug users two years post-service. In *Stress disorders among Vietnam veterans: Theory, research and treatment,* C. Figley, ed. New York: Brunner/Mazel, 1978; Boscarino, J.

References

Current drug involvement among Vietnam and non-Vietnam veterans. *American Journal of Drug and Alcohol Abuse* 6: 301–12, 1979.

9. Musser, M.J., and Stenger, C.A. A medical and social perception of the Vietnam veteran. *Bulletin of the New York Academy of Medicine* 48: 859–69, 1972; Stenger, C.A. The Vietnam veteran. *Psychiatric Opinion* 11: 33–37, 1974.

10. U.S. Congress, House Committee on Veterans Affairs. *Presidential review memorandum on Vietnam-era veterans,* H.R. 38, 10 October 1978.

11. Straker, M. The Vietnam veteran: The task is reintegration. *Diseases of the Nervous System* 37: 75–79, 1976; Lacoursiere, et al. Traumatic neurosis; Friedman, M. Post-Vietnam syndrome: Recognition and management. *Psychosomatics* 22: 931–43, 1981; Penk, W., Robinowitz, R., Roberts, W., Patterson, E., Dolan, M., and Alkins, H. Adjustment differences among male substance abusers varying in degree of combat experience in Vietnam. *Journal of Consulting and Clinical Psychology* 49: 426–37, 1981; Carter, J. Alcoholism in black veterans: Symptoms of posttraumatic stress disorder. *Journal of the National Medical Association* 74: 655–60, 1982; Penk, W., Robinowitz, R., Dolan, M., Gearing, M., and Patterson, E. Interpersonal problems of Vietnam combat veterans with symptoms of posttraumatic stress disorder. *Journal of Abnormal Psychology* 91: 444–50, 1982; Brende, J.O. An educational-therapeutic group for drug and alcohol-abusing combat veterans. *Journal of Contemporary Psychotherapy* 14: 122–36, 1984.

12. Lifton, R.J. *Home from the war.* New York: Simon & Schuster, 1973; Thienes-Hontos, P., Watson, C., and Kucala, T. Stress-disorder symptoms in Vietnam and Korean war veterans. *Journal of Consulting and Clinical Psychology* 50: 558–61, 1982.

11 What Protects Some

1. Cavenar, J.O., and Nash, J.L. Combat and the normal character: War neurosis in Vietnam veterans. *Comprehensive Psychiatry* 17: 648–53, 1976.

2. Hendin, H., Pollinger, A., Singer, P., and Ulman, R. Meanings of combat and the development of posttraumatic stress disorders. *American Journal of Psychiatry* 158: 1490–93, 1981; Hendin, H. Psychotherapy for Vietnam veterans with posttraumatic stress disorders. *American Journal of Psychotherapy* 37: 86–99, 1983.

3. Kasasin, J., Rhode, C., and Wertheimer, E. Observations from a veterans clinic on childhood factors in military adjustment. *American Journal of Orthopsychiatry* 16: 640–59, 1946; Futterman, S., and Pumpian-Mindlin, E. Traumatic war neurosis five years later. *Ameri-*

can *Journal of Psychiatry* 108: 401–8; 1951; Solomon, G., Zarcone, V., Yoerg, R., Scott, N., and Maurer, R. Three psychiatric casualties from Vietnam. *Archives of General Psychiatry* 25: 522–24, 1971; Fox, R. Narcissistic rage and the problem of combat aggression. *Archives of General Psychiatry* 31: 807–11, 1974.

4. Wilson, J.P. *Identity, ideology and crisis: The Vietnam veteran in transition,* Part I. Paper presented to the Disabled American Veterans Association, Forgotten Warriors Project, Cleveland State University, Cleveland, Ohio, 1977; Egendorf, A., Kadushin, C., Laufer, R., Rothbart, G., and Sloan, L. *Legacies of Vietnam: Comparative adjustment of veterans and their peers.* Washington, D.C.: U.S. Government Printing Office, 1981.

5. Hendin, H. Haas, A. Combat adaptations of Vietnam veterans without posttraumatic stress disorders. *American Journal of Psychiatry* 141: 956–60, 1984.

6. Hendin, H., Haas, A., Singer, P., Ulman, R., Gold, F., and Trigos, G. Evaluation of posttraumatic stress in Vietnam veterans. *Journal of Psychiatric Treatment and Evaluation* 5: 303–7, 1983.

7. Caputo, P. *A rumor of war.* New York: Holt, Rinehart & Winston, 1977.

8. Ibid. 4–5.

9. Bloch, H.S. The psychological adjustment of normal people during a year's tour in Vietnam. *Psychiatric Quarterly* 44: 613–26, 1970; Zetzel, E. *The capacity for emotional growth.* New York: International Universities Press, 1970.

12 Conclusion

1. Summers, H.G. *On strategy: A critical analysis of the Vietnam War.* Navato, Ca.: Presidio Press, 1982; Karnow, S. *Vietnam: A history.* New York: Viking Press, 1982.

2. Eisenhart, R.W. "You can't hack it little girl": A discussion of the covert psychological agenda of modern combat training. *Journal of Social Issues* 31: 13–23, 1975.

3. Remarque, E.M. *All quiet on the western front.* Boston: Little, Brown & Co., 1929.

4. Knox, D. *Death march: The suvivors of Bataan.* New York: Harcourt, Brace, and Jovanovich, 1981.

Indexes

Name Index

Name Index

Maxwell, Jerry, 16, 17, 245
May, E., 133, 253
Meyers, Andrew, 108, 109, 185, 187, 252, 254, 255
Menninger, Karl, 25, 246
Mintz, James, 108, 109, 187, 252, 254, 255
Moskos, Charles C., 135, 253
Musser, Marc J., 187, 255

Nace, Edgar, 108, 109, 185, 187, 245, 252, 265
Nail, R., 185, 254
Nash, James, 63, 203, 251, 255
Niederland, William C., 25, 26, 47, 108, 246, 249, 252

O'Brien, Charles, 108, 109, 187, 252, 254, 255
O'Connor, Richard, 18–21, 246
Olson, Eric, 25, 26, 35, 247, 248

Paine, Lewis, 16, 17
Parrish, Matthew D., 7, 136, 245, 253
Parson, Erwin, 89, 160, 251, 254
Patterson, E.T., 188, 255
Penk, Walter, 188, 255
Pumpian-Mindlin, Eugene, 25, 26, 52, 65, 203, 246, 249, 251, 255, 256

Rado, Sandor, 65, 87, 90, 251, 252
Rahe, Richard H., 30, 247, 248

Rappoport, Leon, 133, 134, 253
Ream, Norman, 108, 109, 187, 252, 254, 255
Remarque, Erich Maria, *vii, viii,* 241, 242, 256
Rhode, Charl, 203, 255
Roberts, William R., 188, 255
Robinowitz, Ralph, 188, 255
Robins, Lee N., 106, 186–88, 252, 254
Romano, John, 25, 246
Roscoe, Theodore, 16, 246
Ross, Donald, 26, 247
Rothbart, George, *ix, x,* 7, 204, 245, 256
Rothberg, Joseph M., 185, 254
Ruby, Lorne, 184, 188, 254, 255

Saul, Leon T., 25, 33, 34, 52, 246, 248, 249
Scott, Neil R., 109, 203, 252, 256
Shatan, Chaim, 28, 35, 89, 109, 247, 248, 251, 253
Sherman, Lewis J., 254
Siegel, Arthur J., 184, 185, 254
Simmel, Ernst, 22, 23, 246
Sloan, Lee, *ix, x,* 7, 204, 245, 256
Smith, Clark, 89, 251
Smith, John R., 16, 29, 89, 136, 160, 245, 248, 251, 252, 253
Solberg, Kristen B., 160, 254
Solomon, George F., 109, 203, 252, 253, 256
Sonnenberg, Stephen M., 108, 252
Sparr, Landy F., 62, 89, 250, 251
Spiegel, David, 52, 250
Spiegel, Herbert, 24, 31, 34, 53–54, 58–59, 90, 246, 248, 252
Spiegel, John P., 53, 184, 249, 254
Stenger, Charles A., 187, 255
Sterling, George, 18–21, 246
Straker, M., 188, 255

261

Subject Index

Subject Index

Japanese survivors of atomic bombings, 9, 25–27, 29, 47

Killing, 228; of Americans by Americans in Vietnam, 28, 134–35, 213; fear relieved by, 10, 27; and guilt, 65, 182; lust for, 212–13; non-military, 28, 65, 213–14, 230–31; of prisoners of war, 28, 39–42, 46, 65, 213; random, 239–40; of women, children and the elderly, 4, 28, 65–66, 136–37, 206, 215, 231–32, 238

Lightning neurosis, 21

Marijuana, 71, 96, 167, 185, 195, 197–99
Maternal rejection or loss, 128, 190
Meanings of combat, *see* Combat, meanings of
Medication, 13, 95, 177, 192–93; and therapy, 62–63
Military policy, 8–9, 47–48, 234, 240, 244
Military training, 238–39
Moral ambiguity, 11, 28, 47
Morale, 8, 25, 240
Morphine, 192–93
Mourning, 108–30, 149
Mutilation of bodies, 5, 28, 77–78, 134, 136, 213, 238
My Lai incident, 238

Nagasaki survivors, 9, 25–27, 29, 47
Narcoanalysis, 53–54
Narcotics, 185, 187
Nightmares, 19, 24–26, 34–35; and Freud's views, 23; and the paranoid reaction, 89; as posttraumatic stress disorder symptom, 5, 12, 30; and substance abuse, 13–14, 187, 194–99; and the therapeutic process, 56–57, 61
Non-military violence, 28, 65, 90, 134–36, 213–14, 230–31
Nostalgie, 16

Officers, 210–11, 220, 239–40
Opposition to Vietnam War, 5, 7–8

Paranoid adaptation, 11, 88–107, 129, 235
Political disaffection, 242
Political policy, 47–48, 234, 240, 244
Posttraumatic stress disorder: and criminal behavior, 133–59; definitions of, 22–36; and depressive adaptation, 108–30; in earlier wars, 15–21; and guilt, 65–87; and meanings of combat, 36–48; and paranoid adaptation, 88–107; and pre-existing character and personality, 33–37; prevalence of, 6–7; resistance to, 203–31; social impact of, 7, 241–44; and substance abuse, 183–99; and suicidal be-

265

Subject Index

Traumatic neurosis, 21–24
Treatment, *see* Stress-oriented
  psychotherapy

Veterans Administration, 6, 114,
  160, 187, 197, 236
Vietnam Memorial, 113
Vietnam veterans, age in service,
  x, 8, 233; family background, *x;*
  social class, *x*
Violence, 12–13, 228; and fear of
  retaliation, 107, 136, 233; lack of
  excessive, 212–14; and the paranoid adaptation, 107; postwar,
  90, 133–59; and subtsance abuse,

196–97; and the therapeutic relationship, 60–62; unnecessary,
  28, 65, 90, 213–14, 230–31; *see also*
  Aggression, Killing
Vulnerability, 39, 89, 99, 107,
  209

War neurosis, 34, 53
Women, children, and the elderly, killing of, 4, 28, 65–66,
  136–37, 206, 215, 231–32, 238
World War I, 5, 9, 15, 21–22, 25, 53,
  241
World War II, 5–6, 8–9, 24, 29, 33–
  34, 53, 62, 65, 184, 236, 243